Variational Principles in Physics

Jean-Louis Basdevant

Variational Principles in Physics

Second Edition

 Springer

Jean-Louis Basdevant
Professeur Honoraire
Département de Physique
École Polytechnique
Paris, France

ISBN 978-3-031-21694-7   ISBN 978-3-031-21692-3 (eBook)
https://doi.org/10.1007/978-3-031-21692-3

Original French edition published by de Boeck Superieur S.A., Louvain-la-Neuve, Belgium, 2022
1st edition: © Springer Science+Business Media, LLC 2007
2nd edition: © The Editor(s) (if applicable) and The Author(s), under exclusive license to Springer
Nature Switzerland AG 2023

This Springer imprint is published by the registered company Springer Nature Switzerland AG
The registered company address is: Gewerbestrasse 11, 6330 Cham, Switzerland

Preface

At the Ecole Polytechnique, my major teaching activity was the general course on Quantum Mechanics, but I had many opportunities to get interested in other fields such as Statistical physics, Particle physics, Energy and the environment. The last course I constructed concerned Variational Principles in Physics. Actually, this was unexpected: I had to replace a colleague.

I had not thought about that subject and it was really a discovery that there was much more than what I could teach. But I was lucky enough to have an interesting group of students and colleagues who were happy when I told them we would learn physics together. The spirit was excellent, and together we discovered many aspects of the evolution of physics in the minds of creative people.

In the meantime, two basic results occurred in fundamental physics. First, in July 2012, the discovery, at the CERN Large Hadron Collider, of the long antici-pated Higgs Boson which completed, in some sense, the Standard Model of Particle Physics. On 8 October 2013, the Nobel Prize in Physics was awarded jointly to Francois Englert and Peter Higgs "for the theoretical discovery of a mechanism that contributes to our understanding of the origin of mass of subatomic particles...". A year after, great news occurred with the discovery of the first gravitational waves observed on September 14, 2015, by the LIGO and Virgo laboratories, one hundred years after their prediction by Einstein. These will remain the two most important physical discoveries of the first part of the twenty-first century.

The physics of the Higgs field is too far from the purpose of this book. But it was obviously a must to add a *new* chapter on gravitational waves to the 2007 edition of "Variational Principles in Physics". My colleagues and students said that although variational principles are intellectually attractive, if one does not show what they aim at, they may seem formal.

Another example is the fact that, around 1840, Hamilton was fascinated by an unknown fact: geometrical optics should be considered as a limiting case of wave optics, and classical mechanics seems to be a similar limit of some yet unknown mechanical theory, but which theory? In 1890, the mathematician Felix Klein deplored that no one had pursued that idea. But 35 years later, some of the "fathers" of quantum mechanics, such as Dirac and Louis de Broglie, gained much inspiration from reading Hamilton's works. Furthermore, as we develop and explain in Chap. 9, Richard Feynman in his 1942 Ph.D. dissertation at Princeton, constructed a new version of quantum mechanics: *The principle of least action in quantum mechanics*, based on very simple axioms on probability amplitudes, and on the analogy with Hamilton's "Characteristic function". This is an important progress, in particular, to display in a simple way, analogous to the Huygens–Fresnel principle, how classical mechanics derives as the limit of quantum mechanics. One can follow this in the *new* last chapter, which has been considerably increased, in relation to the *new* Chap. 5 on Hamilton's view of optics and its similarity with mechanics.

A similar remark holds for General Relativity. Many mathematicians of the nineteenth century, such as Gauss, Riemann, and Lobachevsky, were interested in curved spaces. In the Lagrange–Hamilton framework, it is possible to see that the free motion of a particle in a curved space is independent of the particle's mass, so that the equivalence principle of inertial and gravitational mass is achieved, provided the curvature gives a good trajectory. The mathematical work of Einstein and Marcel Grossmann consisted of finding the curvature of space–time that takes into account gravitation.

Laws of nature appear to follow rules expressed mathematically as *variational principles*. These principles possess two characteristics. First, they are universal. Secondly, they express physical laws as the results of optimal equilibrium conditions between conflicting causes. In other words, they present natural phenomena as problems of optimization under constraints. The founding idea of modern physics is due to Fermat and his least time principle in optics. This was further developed in the framework of the calculus of variations of Euler and Lagrange. In 1744, Maupertuis found the least action principle in mechanics.

The philosophical impact of the discovery of such *principles of natural economy* was considerable in the eighteenth century. However, the metaphysical enthusiasm did not last long because variational principles have constantly produced more and more profound physical results, many of which underlie contemporary theoretical physics. The ambition of this book is to describe some of their physical applications. We will see how such ideas have generated most if not all physical concepts of the present time.

After presenting and analyzing some examples, the core of this book is devoted to the analytical mechanics of Lagrange and Hamilton, which is a must in the culture of any physicist of our time. The tools that we develop will also be used to present the principles of Lagrangian field theory.

We then study the motion of a particle in a curved space. This allows us to have a simple but rich taste of general relativity and its first and most famous applications among which gravitational optics allows us to probe the universe at very far distances, as well as the GPS guiding system.

At this point, we can penetrate General Relativity and Einstein's equations by analyzing the production and the reception of the gravitational waves that we have mentioned above and were detected on earth.

In the last chapter, we present a new version of the theory of Feynman path integrals in quantum mechanics. We shall then be able to appreciate how close the Feynman approach is to direct intuitive quantities, space and time, to the ensuing extension of these ideas to a relativistic approach, and to prove that classical mechanics is the limit of quantum mechanics when Planck's constant $\hbar$ is negligible compared to other physical quantities.

I was struck by the interest that students found in this aspect of physics. They discovered a cultural component of science that they did not expect. For that reason, teaching this was a very rewarding piece of work.

I warmly thank Prof. Christoph Kopper, who took over this course at the Ecole Polytechnique. I owe much to David Langlois, Associate Professor who allowed me to study his course and his book on General Relativity. I am extremely grateful to my colleague André Rougé for his useful comments and suggestions.

I am very grateful to James Rich, who was able to extract me from the traditional French academism and make me share his creative enthusiasm for physics. Part of Chap. 7 is directly inspired by his work in a different context.

I thank my colleagues and friends Adel Bilal, François Jacquet, and Jean-François Roussel for all their comments and suggestions when we were teaching this matter and having fun together.

I must thank my mathematician colleagues Jean-Michel Bony, Jean-Pierre Bourguignon and Alain Guichardet. They helped me to avoid spending time on difficult problems that were not necessary.

Finally, I want to thank my students, in particular Claire Biot, Amélie Deslandes, Juan Luis Astray Riveiro, Clarice Aiello Demarchi, Joëlle Barral, Zoé Fournier, Céline Vallot, and Julien Boudet, for their questions and their kind comments. They have provided this book with a flavor and a spirit of youth that would have been absent without them.

Paris, France   Jean-Louis Basdevant
September 2022

Contents

Chapter 1
Structure of Physical Theories

Ancient thinkers and builders were attracted by the optimality of phenomena and their applications. Archimedes studied the optimal form to be given to the hulls of ships, Aristotle, in a more metaphysical mind, claimed that the orbits of planets are circles because their even curvature, which shows neither an origin nor an end, must emanate from a creator who wanted them to be eternal.

The original idea that a physical phenomenon could be optimal is due to Hero of Alexandria,[1] in the first century AD, who explained his experimental observations by a correct geometrical reasoning: in its way between several mirrors, light follows the shortest path.

The variational principles, born with Fermat (1601–1665), Maupertuis (1698–1759), Euler (1707–1783) and Lagrange (1736–1813), marked the continuation of the work of Galileo (1564–1642) and Newton (1642–1727) (and others) towards contemporary physics.

This view of things, quite obvious nowadays, had disappeared from scientific considerations during fifteen centuries, and reappeared in the Variational Principles, which now form the framework of fundamental physical theories, and the Variational Calculus, a particularly fruitful chapter of mathematics. The step forward appears in an astonishing remark of Pierre de Fermat in 1661:

> It is most likely that nature always acts by the easiest means, that is, either by the shortest lines, that do not take more time, or in any case, by the quickest time, in order to reduce its task and to end its operation sooner.

He thus developed a *Method for the search of the maximum and the minimum*,[2] which founded the laws of geometrical optics. This remark coincided with the major rise of

---

[1] Outstanding mathematician and physicist of the first century AD. In particular, he built the first steam engine, the *Aeolipile*.

[2] Fermat's work *Méthode pour la recherche du maximum et du minimum*, was published by Paul Tannery and Charles Henry, Gauthier-Villars, 1896 (Third volume, excerpt, pp. 121–123).

© The Author(s), under exclusive license to Springer Nature Switzerland AG 2023
J.-L. Basdevant, *Variational Principles in Physics*,
https://doi.org/10.1007/978-3-031-21692-3_1

differential calculus in mathematics, and this resulted, a few decades later, in Euler and Lagrange's work on mechanics in 1744. Euler was inspired by Maupertuis's Principle of Least Action. These principles are in the same vein as the observation of Hero of Alexandria.

The list of famous physicists and mathematicians who have since left their names to this work is considerable. This field now forms the conceptual framework of theoretical physics. It inspired the founders of statistical physics and quantum physics, it became an essential tool of quantum field theory and is the basis of general relativity. It is also at the origin of many amazing applications in pure and applied mathematics.[3]

Variational Principles

Variational principles are, in a way, a mathematical form of superlative. This formulation consists in asking that the value of a typical quantity obtained by a system, for the performance actually achieved, be the best compared to what it would be in a different procedure. They are commonly used in applied mathematics.

To some extent, variational principles, through their universality in the world of things, may appear as a general "structure" of physical processes, be it one day in other natural sciences such as biology, psychology or social phenomena. They play a role in all contemporary technologies.

In its first version, a physical theory explains a phenomenon by local laws. Such are the laws of Newton's dynamics, the Descartes-Snell laws, and the differential laws of electromagnetism or thermodynamics. Once the first stone of the theory updated, and after the first holdings, one analyses the underlying principles of the discovery and their links with more general structures.

Variational principles express laws in a *global* form. This form allows, of course, to recover the details of the original local laws, but one discovers that it is deeper, often related to other and more powerful principles. One discovers the fundamental principles that lead to such ideas.

Greek mathematicians characterized a straight segment as the shortest path joining its ends. To say, likewise, that the circle is the shortest line that surrounds a given plane area is an equivalent and more general way of defining this geometric concept.

Similarly, in electricity, saying that the electric current is distributed in a network so that the heat loss by Joule effect be a small as possible is a description that applies to many situations, and can be reduced easily to Kirschhoff's law.

Thus, the variational principles present the natural phenomena as problems of constrained optimization These principles are present in virtually all domains of Physics. (See for instance Chaps. I,26 and II.19 of the Feynman Lectures on Physics [4], and the book of Yourgrau and Mandelstam [6]).

---

[3] The subject is remarkably treated by Jean-Pierre Bourguignon in reference [8].

Fermat and Maupertuis Principles

The mathematical formalisation of such ideas, came first from Pierre de Fermat (1601–1665), as we will see in Chap. 2. His geometric optics principle is a principle of least *time* which gives the law of refraction of Snell-Descartes, and also leads to the understanding of mirages and curved rays.

In fact, everything started around 1637 in a lively critique addressed to Descartes by Fermat about the notion of proof. Fermat's irritation followed the publication of the *Dioptric* in Descartes' *Discourse of the Method*. Fermat, a Toulouse magistrate, was a mathematician, not a physicist, but he was interested in the structure of physical laws.[4] The lack of rigour of the "pseudo-proof" of Descartes, irritated Fermat, who was convinced that things could be done properly: "It seems to me that a little geometry will get us out of this mess". When he succeeded in geometrically demonstrating the law of refraction $n_1 sin\theta_1 = n_2 sin\theta_2$, Fermat was literally fascinated: *[...] I found that my principle gave precisely the same proportion of refractions established by Monsieur Descartes.*

In 1744, Pierre-Louis Moreau de Maupertuis (1698–1759) stated for the first time what he called the *"principle of least amount of action"* for mechanics. He had introduced Newton's ideas in France in 1730. The statement and rationale originally proposed by Maupertuis are very confused, but this is a historic date in the evolution of ideas in physics and, at the time, in philosophy.

He pursued the works of Fermat, and he understood that, in well determined conditions, Newton's equations are equivalent to the fact that a quantity, which he called *Action*, is minimal. In his own words:

> The Action is proportional to the product of the mass by the velocity and by the space. Now, here is this principle, so wise, so worthy of the supreme being: When there is some change in the Nature, the amount of Action used for this change is always as small as possible.

For a particle of mass m, velocity $\mathbf{v}$, the action of Maupertuis is therefore the product of three factors, the mass, the speed, and distance traveled. The formulation and demonstration of Maupertuis's *Principle of Least Action* were, fortunately, given shortly thereafter by Leonard Euler (1707–1783) , his close friend (who realized that Maupertuis's action was the integral of the momentum over the Newtonian trajectory).

These principles had a great impact in the 18th century. That the laws of nature could be deduced from optimization principles, leading to a balance between conflicting causes, struck people's minds. The "principle of natural economy" was a fascinating view of things. It asserted an agreement between different laws of nature in opposition or even in conflict. It was readily linked to the "principle of the best" advocated by Leibniz.

In the last section of Chap. 2, we will turn to a completely analogous case, which is fascinating by the number and the power of its consequences, since it is a complex

---

[4] He had in particular a correspondence with Etienne Pascal, father of Blaise Pascal on mechanical equilibrium.

system, compared to the simplicity of the initial assumption. This is the basis of statistical thermodynamics. By the introduction of the *principle of equiprobability of configurations*, Boltzmann obtains an extremely simple definition of the notion of *temperature*, accompanied by its primary property which is the equalization of temperatures of systems in thermal contact. Then we will arrive at the statistical and absolute definition of *entropy*, due to Boltzmann. This leads to the surprisingly simple principle:

> Thermodynamic equilibrium corresponds to a situation that maximizes entropy, that is, disorder, given the constraints.

This result is the basis of many results in other sciences than physics since the beginning of the 20th century, with the successive developments of the means of communication with electronics and its advances in quantum mechanics, developed vertiginous applications such as the theory of Information, or Communication, by Claude Shannon (1916–2001), an American engineer and mathematician who is its "father", with his founding article *A Mathematical Theory of Communication* published in 1948,[5] which represents a gigantic activity, both scientific and technical, which our societies cannot ignore, and which finds part of its basis in the notion of Boltzmann's Entropy, that we will see in Chap. 2. The "Digital" word becomes an ubiquitous term in today's societies. Quantum Information is a current sector of activity for both researchers and industrial companies.

The same line of thought has been, understandably, one of the cornerstones in the construction of *economic models* . The leading personality in this domain was Paul A. Samuelson (1915–2009) [5], who published his first major work, "Foundations of Economic Analysis"[6] in 1947. His thesis, *Foundations of Economic Analysis* is based on the chemical thermodynamics of Willard Gibbs. He was awarded the (second) Nobel Prize in Economics in 1970 "for the scientific work through which he has developed static and dynamic economic theory and actively contributed to raising the level of analysis in economic science". He was the collaborator of President John. F. Kennedy. The book reveals a common mathematical structure underlying multiple branches of economics from two basic principles: maximizing behavior of agents (such as utility by consumers and profits by firms) and stability of equilibrium such as markets or economies. One of its key insights about comparative statics, called the correspondence principle, states that stability of equilibrium implies testable predictions about how the equilibrium changes when parameters are changed.

The same line of thought has been developed in medicine and biology, in the principles of maximization in evolutionary biology (evolution can be seen as a maximization of entropy[7]), or in the development of population genetics in a Darwinian evolution.

---

[5] Claude E. Shannon, *A Mathematical Theory of Communication*, Bell System Technical Journal, vol 27, no 3, July 1948, pp. 379–423; vol. 27, no. 4, October 1948, pp. 623–666.

[6] Paul A. Samuelson, *Foundations of Economic Analysis* Harvard University Press,1947.

[7] A.W.F. Edwards, Philosophy of Biology, Handbook of the Philosophy of Science.

Lagrange-Euler Analytical Mechanics

Coming back to physics, the variational principles have continued to produce richer and richer physical results.

We will see, in Chap. 3, the contributions of Leonhard Euler (1707–1783) and Joseph-Louis Lagrange (1736–1813), whose work was pursued by William R. Hamilton (1805–1865). These are the fathers, together with Boltzmann, of the cornerstones of contemporary theoretical physics.

The consequences of this vision of physics are at the source of Einstein's general relativity as well as quantum mechanics and modern theories of fundamental interactions.

The central mathematical tool is the *variational calculus*. We owe it to Euler who understood the mechanism, and Lagrange who made a decisive contribution in 1766.[8] Variational calculus is an astonishing part of mathematics, both by its unifying character and by the number of questions it has permitted to answer.

Euler published in 1744 his treatise *Methodus inveniendi lineas curvas maximi minimive Gaudens proprietate*, which founded the variational calculus, that had a considerable influence on Lagrange. It is in this work that Euler justified *a posteriori* the least action principle of his friend Maupertuis.

Lagrange was particularly gifted and precocious. Euler's favourable reaction to his work encouraged him and in 1756 he applied his techniques to the principle of least action in a form that makes it the foundation of modern mechanics and theoretical physics.

One of Lagrange's major contributions is his *Analytic mechanics* where he synthesizes all the methods he had previously developed in statics and dynamics. The work, completed in 1782, appeared only in 1788 in Paris. Lagrange's mechanics is as important in the History of Physic as the celestial mechanics of Newton.

The Role of William Hamilton

This work will be the starting point for all later research, including that of Hamilton, who, given the admiration he had for Lagrange, called it a "scientific poem written by the Shakespeare of mathematics". It was indeed Hamilton who named this theory, and invented the word Lagrangian, after Lagrange.

Chapter 4 will take us to the next step and to the so-called *canonical* formulation of Analytical mechanics, due to Hamilton. It is actually a very long, abundant and deep contribution to physics as well as mathematics (Hamilton invented quaternions, which are the mathematical structure of spin 1/2, which was a terrible problem for physicists who, one century later fought helplessly with this fundamental feature

---

[8] Euler, who had been visually impaired since the age of 28, became blind in the same year 1766. In 1754 he received the visit of the young Lagrange who showed him his works. Amazed by the this young man's talent, he concealed his own results, to leave the sole merit to Lagrange alone. This is an example of an almost unique, and now disappeared, act of human courtesy and passion for science.

of quantum physics during 25 years). This canonical formalism, dating from 1834, is based not on the empirical variables and their time derivatives $(x, \dot{x})$ but on the "conjugated momenta" (x, p). We will see that it brought many people to work on ideas that were completely unpredictable before him and opened new paths to physics. The canonical point of view is more convenient for a number of problems, including mechanics of points or sets of points. But above all, it is impressive in its mathematical and physical developments and its ability to bring out the symmetries of problems.

We will refer to a large number of impacts of Hamilton's work, including some aspects of *dynamical systems*. This type of physical problem has indeed been an extraordinary source of discoveries, both in physics and in mathematics. The founder of this field of study is Henri Poincaré, in 1885, after he studied the three-body problem. This leads to fascinating problems: limiting problems at $t = \infty$, attractors and strange attractors, bifurcations, chaos etc. Perhaps the most famous strange attractor is the Lorenz's attractor, named after its inventor Edward N. Lorenz who discovered it in 1963 from a mathematical model of the atmosphere, and relaunched in a spectacular way, with the "butterfly" effect in meteorology, interest in *chaos*, invented by Poincaré 80 years before.

Hamilton's Link Between Mechanics and Optics

We will then naturally come in Chap. 5, to the amazing discovery of the similarity between analytical mechanics and geometrical optics, which Hamilton had understood.

Between 1825 and 1828 he presented the theory of a single function, the *characteristic function*, which unified mechanics, optics and mathematics and helped him establish the wave theory of light. More precisely, Hamilton showed that mechanics (of that time) had a striking similarity with geometrical optics, which, itself, was the *limit* of wave optics. He wondered about what kind of theory would have classical mechanics as a limit. Of course, at that time there was no experimental manifestation of Planck's constant. But both his characteristic function, which had the dimension of an action, and his remarks were strong sources of inspiration for some fathers of quantum theory, such as Dirac and Louis de Broglie, one century later. We shall come back to this point in the last chapter of this book when we describe Richard Feynman's Path Integral approach to the sources of quantum mechanics.

Hamilton was fascinated by the *action* and the Fermat principle. He understood that it is a *stationary* and not minimal principle of action. The variational principle, also known as the Hamilton principle, is the essential element of his articles. We will describe Hamilton's results on geometrical optics, where he chooses to work directly with *the action* in the form of his characteristic function S, function of canonical variables (x, p). Hamilton formalized the fact that geometric optics is a limit of short wavelength optics, and we will see its amazing structural similarity with mechanics.

Motion in a Curved Space

Einstein's masterpiece is general relativity. That theory is based on the observation that two physical quantities which have *a priori* no relationship, are equal (or strictly proportional). These are, as we know, the two meanings of the concept of mass. One is the coefficient of inertia or resistance to acceleration of a body in the Newtonian laws of dynamics, the other is the coupling coefficient to the field of gravitation. No *a priori* argument explains the rationale of this equality. In a gravitational field, the equality of inertial mass and the heavy mass removes the mass of a body from the equation of motion. By the *Equivalence principle*, two bodies placed in the same initial conditions in a gravitational field have the same motion, regardless of their masses.

The idea behind general relativity is that this equality becomes natural if the motion that we call "gravitational" is, in fact, a *free* motion in a *curved* space-time.

We will see in Chap. 7, the problem of the motion of a free particle in a curved space, with the aim of using the Lagrangian formalism to show how the idea of motion in a curved space provides the desired elements to build a theory where the equality of the "two" masses is achieved naturally.

We will show three historical consequences: the variation of the rate of a clock in a gravitational field, the corrections to the Newtonian celestial mechanics and the deviation of light by the gravitational field. These examples are historical, they are also very topical. Time management has become a daily problem of life in our world (the GPS system). As we shall see, the deflection of light by a field of gravity plays a considerable role in astrophysics and cosmology through the gravitational lens effect. This effect is that of a natural cosmic telescope.

Gravitational Waves

In Chap. 8, we will really address general relativity by describing one of the greatest results in recent years: the quantitative detection of gravitational waves, emitted by accelerated masses. This verification took place one century after Einstein's prediction itself. The first event detected was on September 14, 2015 by the international collaboration LIGO-Virgo. For these results, the 2017 Nobel Prize in Physics was awarded to Rainer Weiss, Barry C. Barish and Kip Thorne. In this case, the waves were emitted by a rotating binary systems of black holes, before they merged into a single congener. Double discovery: gravitational waves and existence of black holes!

Early evidence of gravitational wave emission was recognized by the 1993 Nobel Prize awarded to Joseph H. Taylor and Russell A. Hulse "For the discovery of a new type of pulsar, which opened up new possibilities for the study of gravitation". It was a result of exceptional precision on a totally unexpected phenomenon. Taylor and Hulse had discovered the first example of a double pulsar, on the other hand the rotation of this system emits a gravitational energy so important (although the signal is very weak) that its rotation period decreases over time with an accuracy identical to that of the best theoretical calculations of General Relativity.

Quadrupolar gravitational waves propagate at the speed of light, but they are waves of space-time, that is, of the medium which carries them, and, at the point of detection, their amplitude, measured by the ratio of the relative distance variation (of proper time to be precise) of two points of the detector, is of the order of 10^{-21} (relative order of magnitude of an atom compared to the Earth-Sun distance), and the realization of the detection devices is in itself a prodigy.

Feynman's Variational Quantum Mechanics

Our last Chap. 9 deals with the variational formulation of quantum mechanics due to Feynman.

In 1941, he discovers a 1932 text by Dirac which contains a remarkable idea that will allow him to build a completely new variational formulation of non-relativistic quantum mechanics. In this work, he introduces the mathematical concept of path integrals, which has been developed ever since. This will be the subject of his thesis, defended in May 1942 and published only after the end of the war.

Dirac, like Schrödinger and Louis de Broglie , had reread Hamilton's articles, and in particular meditated on the characteristic function and connection between geometric optics and classical mechanics. He was interested in the phase and ratio of the action and the universal Planck constant $\hbar$, in the expression $exp(iS/\hbar)$, similar to Hamilton's characteristic function in optics. But he did not know how to make an essential calculation.

Feynman solved the problem and formulated quantum mechanics on the basis of the hypothesis of the existence of probability amplitudes, the principle of superposition and this quantum version of the characteristic function. He could thus deduce from these assumptions the form and algebra of observables and the evolution equation, by writing the amplitude of probability of a process as the superposition of amplitudes from the totality of possible quantum paths, generalization of interference by Young's holes to the complete set of possible trajectories. The sum of all the amplitudes realizing the process under consideration is a subtle mathematical object called a *path integral*, on which all the formalism is based.

Feynman showed that one thus obtains the relations of Einstein and Broglie, as well as the Schrödinger equation, the algebra of observables and all the traditional quantum mechanics. Countless results have been achieved, this tool plays a central role in contemporary quantum field theory.

If we consider systems and processes where the classic S action is *macroscopic*, that is to say much greater than the Planck constant $\hbar$, the contribution of paths that may seem very close to each other in the classical sense, but such that the difference in the action calculated on these paths is also, much larger than $\hbar$, will be, with a high probability, in destructive interference. The contribution of all such paths is then zero. In other words, under these conditions, only an infinitesimal neighbourhood of the *classical trajectory*, remains impossible to scrutinize experimentally in its detail. The "probability" of the conventional trajectory is therefore equal to one. Thus, classical mechanics is the limit of quantum mechanics when $\hbar$ is small compared to the quantities of a problem.

This analysis of Richard P. Feynman in 1942 puts a happy end, at least in our present knowledge of physics, to the dream that Willam R. Hamilton had, one century before.

Chapter 2
Variational Principles

Nature always acts by the shortest paths.
Pierre de Fermat

The variational principles present natural phenomena as processes of optimization. The first example concerns the propagation of light rays between several mirrors, as observed by Hero of Alexandria, in the first century, who explained his experimental observations through geometric reasoning: *in its way between several mirrors, the light follows the shortest path.*

In more complex cases, these phenomena result as compromises between the effects of various causes in conflict and become optimizations under constraints. This form of the ideas of Hero of Alexandria reached modern physics only 15 centuries later with Pierre de Fermat (1601–1665) and the principle that we owe him in optics. This vision of things profited considerably from the development at that time of differential calculus, extended to the *variational calculus* developed in mechanics by Leonard Euler (1707–1783) and Joseph-Louis Lagrange (1736–1813) in the 18th century, following the ideas of Pierre-Louis Moreau de Maupertuis (1698–1759) on the "Least Action Principle". The program was accompanied by an important collaboration of remarkable mathematicians, the Bernoulli brothers, Leibniz, Newton, Jacobi, and Hamilton.[1]

In physics, what has been increasingly surprising in this vision of principles is twofold. On the one hand, they present the structures and processes as following from a *principle of optimal conditions*, and, on the other, they are *universal*. All physical laws can be expressed in this global form which, after some analysis, allows to

---

[1] The subject is extensively treated by Jean-Pierre Bourguignon in reference [8].

© The Author(s), under exclusive license to Springer Nature Switzerland AG 2023
J.-L. Basdevant, *Variational Principles in Physics*,
https://doi.org/10.1007/978-3-031-21692-3_2

recover and find local laws such as Newton's law on the proportionality of force and acceleration of a body in mechanics, or Coulomb's law on the force between two electric charges. The depth of this approach, more rich and more powerful, give their full origin to local laws, and exhibit the fundamental principles which are behind them. In fundamental physics, quantum physics, elementary particles or general relativity, it is mainly symmetry properties and invariance laws that govern the construction of theories.

In this chapter, we review some important examples and introduce the necessary mathematical tools that show how they work. In Sect. (2.1), we turn back to the Fermat principle, in particular Fermat's proof of the laws of refraction. Fermat did not know the velocity of light and the existence of an index of refraction. He assumed that the time it takes light to travel a certain distance in a medium is increased by the "resistance" of that medium to the propagation of light. Fermat stated his "least time principle" at the end of 1661. He called it the "principle of natural economy." This principle explains curved light rays and mirages, which the Snell–Descartes laws do not account for. This will directly lead us to the central underlying mathematical foundation of the problem under consideration: the variational calculus of Euler and Lagrange in Sect. (2.2). It is an amazing chapter of mathematics, both by its unifying aspect and in the number of problems that it allows to solve. The mathematical details are remarkably exposed by Jean-Pierre Bourguignon [8]. It is deliberately that we shall not go into any mathematical details. Such details can be found in the literature, and we shall focus on physical applications and results[2]

The direct consequences of this first mathematical approach are the phenomena of optical mirages that we discuss and illustrate in section and, In Sect. (2.3), we will give first example of the "least action principle," as first stated by Maupertuis in mechanics in 1744. This was a landmark in the evolution of ideas in physics as well as philosophy at the time.

In this chapter, we will examine three important examples and the mathematical tools needed to formalise them. In Sect. 2.1, we return to the Fermat principle, and in particular the latters demonstration of the laws of reflection and refraction. Fermat didnt know about the speed light and refractive indices. Assuming that the time it takes light to travel in a medium is increased, Fermat enunciated at the end of 1661 his *least time principle* that he called the "principle of natural economy". This principle directly explains, as we shall see, the *curved rays*, responsible for mirage phenomena, which are absent in the framework of the Snell-Descartes laws.

This will lead us, in Sect. (2.2), to a first approach to the mathematical core of our subject: the *variational calculation* of Euler and Lagrange. This is an amazing part of mathematics, both by its unifying aspect, by the number of questions it allows to be answered and by the diversity of its consequences and applications. Well come back to those points later on. The mathematical details are remarkably exposed by

---

[2] In addition to this mathematical presentation, a classic treatise is C. Lanczos book, *The Variational Principles of Mechanics* Dover Publications, 1970, and Lawrence Schulmans highly detailed work [27]; there is a basic presentation in: P.K. Townsend *Variational Principles, Part 1A Mathematics Tripos* 2018, https://www.damtp.cam.ac.uk/user/examples/B6La.pdf.

Jean-Pierre Bourguignon [8] and by H. Goldstein [3]. It is deliberately that we have not addressed these mathematical aspects, which have been extensively discussed in the literature[3]

Next, we will see in Sect. (2.3) how the *least action principle of* Maupertuis for the mechanics of the point, allows to find Newton's law. This marks a historical date in the evolution of ideas in physics.

In Sect. (2.4), we turn to a case of a complex nature, but completely analogous in spirit, which fascinates by the quantity and power of its consequences, compared to the simplicity of the initial hypothesis. This is the foundation of thermodynamics by Ludwig Boltzmann (1844–1906). Erwin Schrödinger (1887–1961) wrote [9] that there is only one problem in statistical thermodynamics: to determine the distribution of a given quantity of energy E over N identical systems. By introducing the *principle of equiprobability of configurations* and the technique of Lagrange multipliers, we will see how emerges a very simple definition of the notion of temperature, a concept that has long been obscure, in contrast to that of heat, accompanied by its primary property, which is the equalization of temperatures of two systems in thermal contact at equilibrium. One then arrives at the statistical and absolute definition of *entropy*, due to Ludwig Boltzmann. This leads to the surprisingly simple principle:

> *The thermodynamic equilibrium of a complex system corresponds to a a situation that maximizes entropy, that is, its* disorder, *given its* constraints.

Finally, in Sect. (2.5), we will go over some other simple and classic examples.

2.1 Fermat's Least Time Principle

As we have already mentioned, everything started with a quarrel between Descartes and Fermat in 1637 on the notion of proof after the publication of "Dioptrique" in Descartes's *Discours de la Méthode*.

The Snell–Descartes laws predict which path a given initial light ray will follow. Fermat takes a more general point of view. He wants to determine what path a light ray *actually follows* when it goes from A to B. We know that this point of view allows us to explain curved light rays and mirages, whereas the Snell–Descartes laws cannot do so. Fermat understands, as did Hero of Alexandria, that the law of reflection is a geometric property of the optical length of the light rays. The proof is sketched in Fig. 2.1.

Consider an emitter A and an observer B. We assume that light emitted by A is reflected by a plane mirror before it reaches B. Let B' be the symmetric to B with respect to the plane of the mirror, and O the intersection of the mirror and the straight line AB' (Fig. 2.1). The length of AOB' is the same as that of AOB. The shortest

---

[3] In addition to this mathematical presentation, a classic treatise is the book of C. Lanczos *The Variational Principles of Mechanics*, Dover Publications, 1970, and Lawrence Schulmans highly detailed work [27]; there is a basic presentation in: P.K. Townsend *Variational Principles, Part 1A Mathematics Tripos* 2018, https://www.damtp.cam.ac.uk/user/examples/B6La.pdf.

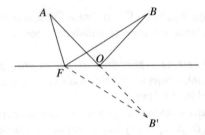

Fig. 2.1 Possible light rays between an emitter A and an observer B when there is a reflection on a plane. Since B' is symmetric to B with respect to the plane of the mirror the length of AOB' is equal to that of AOB. The shortest path between A and B' is a straight line. A path AFB is longer whenever $F \neq O$

distance between A and B' is quite obviously a straight line. A path AFB where $F \neq O$ is such that by the triangle inequality (or by the definition of a straight line) $AF + FB' > AB'$, whatever F. Elementary geometry then shows that the angles of incidence i and reflection r are equal for the path AOB.

Refraction

Concerning the laws of refraction, Descartes had assumed that the velocity of light in matter (a dense medium) is *larger* than in vacuum (or in a diluted medium).[4] That fact, together with the lack of rigor of Descartes's "proof," had made Fermat angry. He was convinced that things could be done properly.

Fermat solved the problem of refraction only much later, in 1661, annoyed by the critiques of Descartes's supporters. The key point of his proof lies in the assumption that the velocity of light is just *smaller* in a dense medium than in a dilute one.

Let (X, Y) be the plane separating the two media of indices n_1 and n_2. The source is at point A and the observer is at point B, as represented in Fig. 2.2.

Let H and H' be the projections of A and B on the (x, y) plane. We denote by h the distance of A to the surface and h' that of B. The distance HH' is l. We consider a path AOB and we denote by x the distance HO. We want to minimize the optical path $n_1 AO + n_2 OB$.

By the Pythagorean theorem, we have

$$AO^2 = h^2 + x^2, \quad OB^2 = h'^2 + (l - x)^2.$$

The time T it takes light to follow this path is

$$T = (n_1 AO + n_2 OB)/c. \tag{2.1}$$

---

[4] This idea probably comes from the fact that many interfaces under consideration were horizontal liquid surfaces, perpendicular to the direction of gravity. Since the refracted light ray appears to be closer to the vertical when it passes from air to water, for instance, it seemed intuitive that it "falls" more rapidly.

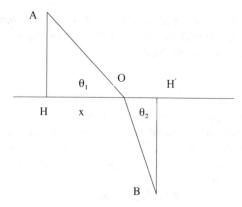

Fig. 2.2 Possible light ray between an emitter A and an observer B when there is refraction across a plane surface between two media of indices n_1 and n_2. H and H' are the projections of A and B on the surface. h is the distance between A and the surface, and h' that between B and the surface. The distance HH' is l

Here, we give an analytic proof, at present simpler to understand than the beautiful purely geometric argument of Fermat. (Fermat did not fully know differential calculus, which was developed later by Newton and Leibniz, although he had correct ideas on the subject.)

We seek x such that (2.1) is minimal. By taking the derivative of this expression with respect to x, and writing that the derivative dT/dx vanishes, we obtain

$$\frac{n_1 x}{\sqrt{h^2 + x^2}} = \frac{n_2 (l - x)}{\sqrt{h'^2 + (l - x)^2}}. \tag{2.2}$$

We note that

$$\frac{x}{\sqrt{h^2 + x^2}} = \cos \theta_1 = \sin i_1, \quad \text{and} \quad \frac{(l - x)}{\sqrt{h'^2 + (l - x)^2}} = \cos \theta_2 = \sin i_2, \tag{2.3}$$

where the angles θ_1 and θ_2 are indicated In the figure, and i_1 and i_2 are the angles of incidence and refraction.

Therefore, we obtain the Snell–Descartes law

$$n_1 \sin i_1 = n_2 \sin i_2. \tag{2.4}$$

Furthermore, this extremum is indeed a *minimum* ($d^2T/dx^2 > 0$).

This result fascinated Fermat: "The outcome of my work was the most extraordinary, the most unexpected and the happiest that ever was. Indeed ... I found that my principle gave exactly and precisely the same proportion of refractions as Mr. Descartes has found." At the end of 1661, Fermat wrote his principle of least time, which was the first formulation of everything that concerns us. Fermat called it the "principle of natural economy," and he added the remark that *"Nature always acts by the shortest paths."* As we said, this principle had a great impact in the 18th century. It was used by Maupertuis in mechanics.

Rescuing a Swimmer

This result can be transposed into many other situations. One example is the optimal path that a rescuer must follow on a beach and in the water in order to rescue a bather in difficulty as quickly as possible. The velocities of the rescuer on the beach, v_1, and in the water, v_2, are not the same. The optimal trajectory, which can be sketched as in (2.2), obeys the law

$$\sin i_1 / v_1 = \sin i_2 / v_2.$$

Curved Rays

Consider a two-dimensional problem (x, z) such as the propagation of light in a fixed atmosphere whose density varies so that the index of refraction varies continuously from one point to another. The situation is represented in Fig. 2.3.

The light rays propagate along curved paths and not straight lines, and the optical angular position of an object differs from its geometrical direction.

From the mathematical point of view, we need to find the path $z = Z(x)$ of a light ray propagating in a medium of index $n(z, x)$ (or simply $n(z)$ if the system is translation invariant along the x direction) and going from a point A at (z_0, x_0) to an observer B at (z_1, x_1). The time $d\tau$ that it takes light to go from $[x, z]$ to $[x + dx, z + dz]$ is

$$d\tau = n(z)\frac{dl}{c} = n(z)\frac{\sqrt{dz^2 + dx^2}}{c}.$$

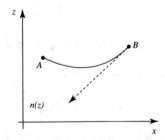

Fig. 2.3 Light ray between an emitter A and an observer B in a medium whose index of refraction varies with the altitude z. The variable x is the horizontal distance. We assume that the problem is translation invariant in the perpendicular y direction. The apparent direction of point A as seen by B is the tangent to the light ray reaching B

By definition, along the path $Z(x)$ that we wish to determine, we have $dz = (dZ(x)/dx)\, dx \equiv \dot{Z}(x)dx$.

Therefore, we must find the function $Z(x)$ that minimizes the *time along the path*; in other words, the integral

$$T = \frac{1}{c}\int_A^B n(z)\sqrt{1 + \dot{z}(x)^2}\, dx, \tag{2.5}$$

given the constraints $Z(x = x_0) = z_0$ and $Z(x = x_1) = z_1$.

This leads us to the mathematical core of this subject.

2.2   Variational Calculus of Euler and Lagrange

The problem under consideration consists in finding a function, or a family of functions, that minimizes some integral. It is called the *variational calculus*. This part of mathematics was developed by Euler, who understood how it functions, with the decisive contributions of Lagrange.

The variational calculus is an amazing chapter in mathematics. It bears many unifying features and it gives the answers to a large number of questions. It is fully treated in the literature, and we do not wish to enter into mathematical details in this book; all details can be found in the works listed in reference [8].

The elementary problem is the following. Find the function $z(x)$ of a real variable x that minimizes (or maximizes) the integral

$$I = \int_A^B \mathcal{L}(z(x), \dot{z}(x), x)dx. \tag{2.6}$$

where the endpoints A and B are fixed, where $\dot{z}(x) \equiv dz(x)/dx$, and where $\mathcal{L}$ is a known function, called the Lagrange function. Needless to emphasize, it is exactly the problem of Eq. 2.5.

Let us assume there exists a solution, that we denote as $z = Z(x)$. We want that, for any infinitesimal variation $\delta z(x)$ of $Z(x)$, there corresponds a *second-order* (or more) variation of the integral I. In the transformation $Z \to Z + \delta z(x)$, $\dot{Z} \to \dot{Z} + \delta \dot{z}(x)$, where it is assumed that the endpoints of the integration do not change, $\delta z(A) = \delta z(B) = 0$, the variation δI of the integral is

$$\delta I = \int_A^B \left[\frac{\partial \mathcal{L}}{\partial z}\delta z(x) + \frac{\partial \mathcal{L}}{\partial \dot{z}}\delta \dot{z}(x) \right] dx.$$

The second term can be integrated by parts since by definition $\delta \dot{z} = (d/dx)\delta z$. Since $\delta z(A) = \delta z(B) = 0$, the variation δI is

$$\delta I = \int_A^B \left[\frac{\partial \mathcal{L}}{\partial z} - \frac{d}{dx}\left(\frac{\partial \mathcal{L}}{\partial \dot{z}} \right) \right] \delta z(x)\, dx. \tag{2.7}$$

We want this integral to vanish for any infinitesimal variation δz. The integrand must vanish identically. Therefore, the solution $z = Z(x)$ must satisfy the second-order differential equation

$$\frac{\partial \mathcal{L}}{\partial z} = \frac{d}{dx}\left(\frac{\partial \mathcal{L}}{\partial \dot{z}} \right), \tag{2.8}$$

called a Lagrange–Euler equation .

This procedure can be extended to a function $\mathcal{L}(\{z_i(x), \dot{z}_i(x)\}, x)$ of several variables $z_i(x)$, $i = 1, \ldots, N$. This leads to N Lagrange–Euler equations

$$\frac{\partial \mathcal{L}}{\partial z_i} = \frac{d}{dx}\left(\frac{\partial \mathcal{L}}{\partial \dot{z}_i} \right). \tag{2.9}$$

2.2.1 First Integrals, Cyclic Variables

The Lagrange-Euler equation (2.8) is a differential equation of the second order. In some cases, we can go back to solving a first-order equation. We will return to this question in Chap. 3, when we discuss symmetries and conservation laws. It is convenient here to consider three simple cases where Lagrangian does not explicitly depend on one of the variables.

We then arrive at an equation of first order called a *first integral* of the problem under consideration.

1. $\mathcal{L}$ does not depend on z. The Eq. (2.8) is reduced to

$$\frac{\partial \mathcal{L}}{\partial \dot{z}} = C_1. \tag{2.10}$$

where C_1 is a constant. This equation is easily solved.
2. $\mathcal{L}$ does not depend on $\dot{z}$. Equation (2.8) then reduces to

$$\frac{\partial \mathcal{L}}{\partial z} = 0. \tag{2.11}$$

which implicitly defines $z(x)$.
3. The case when $\mathcal{L}$ does not depend on x is frequent and interesting. We then have $(d/dx)(\mathcal{L}) = 0$ that we carry in the expression

$$\frac{d\mathcal{L}}{dx} = \frac{\partial \mathcal{L}}{\partial x} + \dot{z}\frac{\partial \mathcal{L}}{\partial z} + \ddot{z}\frac{\partial \mathcal{L}}{\partial \dot{z}}.$$

This can be rewritten in the form:

$$\frac{d\mathcal{L}}{dx} = \dot{z}\left[\frac{\partial \mathcal{L}}{\partial z} - \frac{d}{dx}\left(\frac{\partial \mathcal{L}}{\partial \dot{z}}\right)\right] + \frac{d}{dx}\left(\dot{z}\frac{\partial \mathcal{L}}{\partial \dot{z}}\right),$$ (2.12)

or equivalently:

$$\frac{d}{dx}\left(\mathcal{L} - \dot{z}\frac{\partial \mathcal{L}}{\partial \dot{z}}\right) = \dot{z}\left[\frac{\partial \mathcal{L}}{\partial z} - \frac{d}{dx}\left(\frac{\partial \mathcal{L}}{\partial \dot{z}}\right)\right].$$ (2.13)

Taking into account the Lagrange-Euler equation (2.8), we obtain

$$\frac{d}{dx}\left(\mathcal{L} - \dot{z}\frac{\partial \mathcal{L}}{\partial \dot{z}}\right) = 0 \quad \Rightarrow \quad \mathcal{L} - \dot{z}\frac{\partial \mathcal{L}}{\partial \dot{z}} = C$$ (2.14)

where C is a constant. This is a first-order differential equation that will help us solve some problems in a simple way.

We shall see, in Chap. 3, that in mechanics, the fact that the lagrangian does not depend explicitly on time, and is therefore invariant by translation in time, implies the fundamental law of conservation of energy.

2.2.2 Mirages and Curved Rays

Let us come back to the case of curved rays considered above. Consider the integral (2.5) and let us assume for definiteness that the index of refraction varies with the altitude as $n(z) = 1 + \nu z$, with ν constant. We also assume that the endpoints are at the same height $z(x = 0) = h$ and $z(x = l) = h$. The Lagrange function is

$$\mathcal{L} = \frac{1}{c}(1 + \nu z)\sqrt{1 + \dot{z}(x)^2},$$

from which we deduce the Lagrange-Euler equation

$$\nu(1 + \dot{z}^2) = (1 + \nu z)\ddot{z}.$$ (2.15)

The solution of this equation is simple. We shift to the function $u = z + 1/\nu$ and insert this into (2.15). We obtain

$$1 + \dot{u}^2 = u\ddot{u},$$ (2.16)

whose general solution is

$$u = d\cosh((x - b)/d),$$ (2.17)

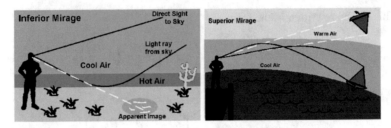

Fig. 2.4 Diagrams of an inferior mirage (left) and a superior mirage (right)

where b and d are constants.

One way to obtain this result consists of taking the derivative of (2.16). This yields $(u'/u) = (u'''/u'')$, whose "solution" is $u'' = u/d^2$, where d is an arbitrary constant. The solution of this latter equation is $u = a \cosh((x - b)/d)$, where, if we use (2.16), $a = d$.

One can obtain the result (2.17) in a more elegant fashion by using the conservation laws associated with the Lagrange–Euler equations, as we shall see in Chap. 3.

If we impose the boundary conditions (i.e., constraints) $z(0) = h$ and $z(l) = h$, we obtain the result

$$z = d \cosh((x - l/2)/d) - 1/\nu \quad \text{with} \quad d \cosh(l/2d) - 1/\nu = h. \qquad (2.18)$$

In this simplistic model, the trajectory of the curved ray is a cosh function whose minimum (or maximum) altitude is attained at $x = l/2$ (the symmetry of the problem).

This situation is encountered in mirages. Perhaps the most common is highway shimmer. Parts of a hot highway can appear as "lakes." This sort of mirage is sketched in Fig. 2.4. The index of refraction is smaller near the highway, where the temperature is high and the air less dense, whereas it increases with the altitude, where the temperature is lower. The "lake" is a reflection of the sky. Such a case is called an *inferior mirage*. The apparent image is below the actual direction of the object.

As one can understand from this simple example, a more complex variation of the index of refraction $n(z)$ will lead to a variety of phenomena. The reverse happens if the index is smaller at high altitudes than at lower altitudes. This type of situation happens when light rays propagate near a hot hill. These are called *superior mirages*. One can then see an object that should be hidden geometrically by the hill, such as the mirages in the desert.

At sunset, one can see the sun for quite a long time after it has gone below the geometrical horizon. As shown in Fig. 2.5, when the sun is close to the horizon, light rays cross an atmosphere whose index of refraction varies considerably with the altitude. At sunset, the angle between the apparent direction of the sun and its actual direction is roughly half a degree. The sun is far below the horizon (see [4] for other examples).

Fig. 2.5 Actual and apparent directions of the sun near the horizon. They differ by ~0.5 degrees

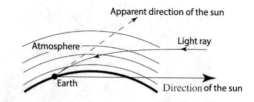

Fig. 2.6 Double superior mirages observed by sailors of the *Germania* expedition in the Arctic in 1888. (Courtesy of Roger Lapthorn.)

Mirages happen frequently in the Arctic and Antarctic. For a long time, the line of sight crosses a large thickness of the atmosphere. Over that distance, the density and chemical composition of the atmosphere can vary considerably. This results in spectacular effects.

Figure 2.6 is a picture taken during a German expedition led by the ship *Germania* in the Arctic in 1888. It is particularly rich, since for both ships there are two superior mirages, inverted with respect to one another. Between the ships, one can see an iceberg. This picture is reminiscent of the legend of the *Flying Dutchman*[5] (in the Southern Hemisphere).

Figure 2.7 shows two mirages. Above, an inferior, inferior mirage of the sun in the forefront is caused by the strong density and temperature variations inside a layer of clouds visible on the picture Below superior mirage of icebergs in the Arctic (Courtesy of Pekka Parviainen.). See also http://www.atoptics.co.uk/.

The variations of the index of refraction of the atmosphere generate a series of effects, in particular lensing effects. It is possible to observe islands, ships, and coasts that are several hundred kilometers away. The variation of the index allows one to see and take pictures of the famous "Green ray" at sunset (see Pekka Parviainen at http://virtual.finland.fi/finfo/english/mirage2.html).

---

[5] The "Flying Dutchman" was a famous sailor. He claimed he could sail around the Cape of Good Hope whatever the weather conditions. Years after he disappeared in a huge storm, many sailors claimed they had seen his ship, in particular in the sky, which was proof that storms were unable to beat him.

Fig. 2.7 Above: remarkable double sunset mirage observed at Paranal Observatory in the Atacama Desert, Chile, by Luc Arnold in 2002 at the site of the European Southern Observatory Very Large Telescope. (Courtesy of Luc Arnold). Below superior mirage of icebergs in the Arctic. (Courtesy of Pekka Parviainen.)

2.3 Maupertuis, Principle of Least Action

In 1744, Maupertuis stated for the first time his *principle of the least quantity of action* in mechanics. Even though the initial version and justification of Maupertuis are confused, it is a historical landmark in the evolution of ideas, both in physics and, at the time, in philosophy.

Consider a particle of mass m and velocity v. The action of Maupertuis is the product of three terms: the mass, the velocity, and the distance covered. Actually, it is the integral of the linear momentum along the trajectory: $A = \int mv\, dl$.

The correct formulation and the proof of Maupertuis's principle were given a little later by Euler. In present terminology, consider a point-like particle of mass m in a potential $V(\mathbf{r})$. We denote by $\mathbf{v}$ the velocity and v its norm. Assuming (this is essential) that the energy E is a constant of the motion, we have

$$E = \frac{1}{2}mv^2 + V(\mathbf{r}).$$

The action of Maupertuis is

$$A_{a,b} = \int_a^b mv \, dl \equiv \int_a^b \sqrt{2m(E-V)} \, dl, \tag{2.19}$$

where dl is the length element along the trajectory. The principle of Maupertuis is that the physical trajectory that the particle follows to go from a to b with a fixed energy E is the path that makes (2.19) minimum.

There are many proofs. We parameterize the state variables $\{\mathbf{r}, \dot{\mathbf{r}}\}$, where $\mathbf{r} = (x, y, z)$, by the time t *on the physical trajectory* (i.e., we work with $\{\mathbf{r}(t), \dot{\mathbf{r}}(t)\}$). The times of departure t_a and arrival t_b are therefore well defined. We have $dl = v \, dt = \sqrt{(\dot{x}^2 + \dot{y}^2 + \dot{z}^2)} \, dt$, and the action (2.19) is

$$A_{a,b} = \int_{t_a}^{t_b} \sqrt{2m(E-V(\mathbf{r}))} \, \sqrt{\dot{x}^2 + \dot{y}^2 + \dot{z}^2} \, dt. \tag{2.20}$$

We want this quantity to be stationary under infinitesimal variations $\mathbf{r} \to \mathbf{r} + \delta \mathbf{r}$, $\dot{\mathbf{r}} \to \dot{\mathbf{r}} + \delta \dot{\mathbf{r}}$. The Lagrange–Euler equations (2.9) lead to

$$- mv \frac{\nabla V}{\sqrt{2m(E-V(\mathbf{r}))}} = \frac{d}{dt} \left(\frac{\dot{\mathbf{r}}}{v} \sqrt{2m(E-V(\mathbf{r}))} \right) \tag{2.21}$$

By definition, along the trajectory that we are looking for, we have

$$2m(E-V(\mathbf{r})) = m^2 v^2.$$

Therefore, Eq. (2.21) boil down to

$$- \nabla V = m \frac{d\mathbf{v}}{dt} \qquad QED. \tag{2.22}$$

2.3.1 *Electrostatic Potential*

Consider now a slightly more complicated problem. We want to determine the electrostatic potential $\phi(\mathbf{r})$ created by a given distribution of charges $\rho(\mathbf{r})$. We know that the answer is Poisson's law,

$$\Delta \phi = - \frac{\rho}{\varepsilon_0}. \tag{2.23}$$

This result can be obtained by the following variational principle (which is a particular case of a more general principle concerning Maxwell's equation, as we shall see in Chap. 6). The electrostatic field is expressed in terms of the potential by

$\mathbf{E} = -\nabla\phi$, and the field energy is $\mathcal{E}_E = (\varepsilon_0/2) \int \mathbf{E}^2 \, d^3\mathbf{r}$. The electrostatic potential energy of a charge distribution $\rho(\mathbf{r})$ in the potential $\phi(\mathbf{r})$ is $\mathcal{E}_\rho = \int \rho(\mathbf{r})\phi(\mathbf{r}) \, d^3\mathbf{r}$.

The variational principle here is that the physical potential $\phi(\mathbf{r})$ minimizes the *difference* between these two energies (or maximizes it if one takes the opposite expression). Consider the integral

$$U = \int [(\varepsilon_0/2)(\nabla\phi)^2 - \rho(\mathbf{r})\phi] \, d^3\mathbf{r}. \tag{2.24}$$

The problem under consideration is to find the potential $\phi(\mathbf{r})$ that minimizes this expression.

We remark on the following points:

1. As usual, we assume there are no charges at infinity, so that ϕ can be chosen to vanish at infinity. The integrals run over all three-dimensional space.
2. Since the first term is positive, if there exists a minimum of this expression for a function $\phi(\mathbf{r})$, this minimum corresponds to an *equilibrium* situation. In this respect, it is similar to the case of the massive string in Sect. 2.5. There is an equilibrium between two contributions to the total energy that *compete* with one another. Any *excess* of one form of energy corresponds to an unstable situation.
3. In comparison with the mirage (2.5) or the massive string (A.1), it is the potential ϕ and its gradient $\nabla\phi$ that play the role of the previous single variable z and its derivative $\dot{z}$. The variable x of the previous simple examples is now a point $\mathbf{r}$ of three-dimensional space (i.e., $\mathbf{r} \in \mathcal{R}^3$).

Let ϕ be the solution and $\eta(\mathbf{r})$ an infinitesimal variation of this potential. In the variation $\phi \to \phi + \eta$, we have, to first order,

$$(\nabla\phi)^2 \to (\nabla\phi)^2 + 2\nabla\phi \cdot \nabla\eta.$$

Therefore, the variation of (2.24) is

$$\delta U = \int [\varepsilon_0(\nabla\phi \cdot \nabla\eta) - \rho\eta] \, d^3\mathbf{r}. \tag{2.25}$$

Integrating the first term by parts, and taking into account the fact that ϕ vanishes at infinity, we obtain

$$\int (\nabla\phi \cdot \nabla\eta) \, d^3\mathbf{r} = -\int \Delta\phi \, \eta \, d^3\mathbf{r}.$$

Therefore,

$$\delta U = \int [-\varepsilon_0\Delta\phi - \rho] \, \eta \, d^3\mathbf{r}. \tag{2.26}$$

The fact that $\delta U = 0$ for any infinitesimal $\eta(\mathbf{r})$ yields the Poisson Eq. (2.23)

$$\Delta\phi = -\frac{\rho}{\varepsilon_0}.$$

A particular case is when the charge density vanishes. By that, we mean that there are a certain number of charged conductors each of which is at a given potential $V_1, V_2, \ldots, V_n$. There is a surface charge density, but the volume density ρ vanishes everywhere. Let $\Sigma_1, \Sigma_2, \ldots, \Sigma_n$ be the surfaces of the conductors. Then Eq. (2.26) boils down to

$$\Delta\phi = 0,$$

with the n constraints $\phi = V_i$ on Σ_i.

2.4 Thermodynamic Equilibrium: Maximal Disorder

2.4.1 Principle of Equal Probability of States

Let us turn to a case that is similar in its motivation but that has fascinating consequences compared with the simplicity of the starting point.

As Schrödinger wrote [9], there is, basically, only one problem in statistical thermodynamics: the distribution of a given amount of energy E over N identical systems.

We only consider here *classical* statistical thermodynamics. Quantum statistics is outside the scope of this book. The only "quantum" feature lies in the fact that we assume there are discrete energy levels.

We consider an isolated assembly of N identical systems $\{s_1, s_2, \ldots, s_N\}$, each of which can occupy one of the energy levels ε_k (for instance, the energy levels in a box where we place the atoms of a monatomic gas).

We assume that the pairwise interactions of these systems are weak in the sense that they do not affect their energy levels. The energy of the assembly is therefore the sum of the energies of the N systems.

Let us call the *state* or *configuration* of the assembly the fact that

system s_1 has energy e_1,
system s_2 has energy e_2,
system s_3 has energy e_3, etc.

The e_i belong to the set $\{\varepsilon_k\}$ and, of course, the sum is equal to the (given) total energy E.

We call *distribution* of the N systems the fact that

n_1 systems are in the energy level ε_1,
n_2 systems are in the energy level ε_2,
n_3 systems are in the energy level ε_3,

etc.

with the conditions or constraints

$$\sum_i n_i = N \quad \text{and} \quad \sum_i n_i \varepsilon_i = E.$$

Of course, for a given distribution, there correspond several states or configurations. Their number W is

$$W = \frac{N!}{n_1! n_2! \cdots n_k! \cdots}. \tag{2.27}$$

The fundamental premise of statistical physics is extremely simple. It is called the *principle of equal probability of states* or *configurations*.

All states, or configurations, of an isolated assembly of systems in weak mutual interaction with total energy E are equally probable.

In other words, if it were possible to take pictures, at given times, of the assembly, or its state, one would observe that the probability of finding the assembly in any one of its possible states is the same.

The consequence of this assumption is that the probability of finding the assembly in a given *distribution* $\{n_1 \, n_2 \ldots, n_k, \ldots\}$ is proportional to W (2.27).

2.4.2   Most Probable Distribution and Equilibrium

In a macroscopic assembly, the number of systems is extremely large. Among all the $\{n_i\}$ possible distributions, there is one in the vicinity of which the number W is maximum. Furthermore, and this can be proven, W has a sharp maximum in the vicinity of that distribution. This particular vicinity is *much more probable* than any other distribution. In other words, if one were to inspect the state of the assembly, one would most of the time find a state in the vicinity of the most probable distribution.

This distribution (more correctly, this vicinity) corresponds to the *thermodynamic equilibrium* of the assembly.

We therefore want to determine the distribution that *maximizes W*. Actually, W is a very large number. It is convenient to maximize its logarithm rather than W itself.

Since the numbers $\{n_i\}$ are very large, we can use Stirling's formula $N! \sim N^N e^{-N} (2\pi N)^{1/2}$ (where the last factor doesn't play any significant role), which leads to

$$\ln W = N \ln N - N - \sum_i n_i \ln n_i + \sum_i n_i = \sum_i n_i \ln(N/n_i) = -N \sum_i p_i \ln(p_i), \tag{2.28}$$

where we have introduced the probabilities $p_i = n_i/N$.

We want to find the distribution $\{n_i\}$ that maximizes this expression under the constraints

$$\sum_i n_i = N, \quad \sum_i n_i \varepsilon_i = E. \tag{2.29}$$

2.4.3 Lagrange Multipliers

In order to solve this problem, we use the technique of Lagrange multipliers, which has many applications.

The problem under consideration is to find the maximum of a function $f(x, y)$ with the *constraint* that (x, y) lie on a path $y = y_0(x)$ (for instance, find the highest point not of a mountain but of a road on that mountain).

Of course, one can think of injecting the equation of the path in f and calculating x such that

$$\frac{d}{dx} f(x, y_0(x)) = \frac{\partial f}{\partial x} + \frac{\partial f}{\partial y} \frac{d}{dx}(y_0(x)) = 0. \tag{2.30}$$

The method of Lagrange has several practical advantages. It consists of introducing an auxiliary function $g(x, y) = y_0(x) - y$ and a *new variable* λ, called a Lagrange multiplier. We search for the extremum of the function of *three* variables (x, y, λ),

$$f(x, y) + \lambda g(x, y) \quad \text{with} \quad (x, y) = y_0(x) - y. \tag{2.31}$$

We must therefore solve three equations:

$$\frac{\partial f}{\partial x} + \lambda \frac{\partial g}{\partial x} = 0 \ (1), \quad \frac{\partial f}{\partial y} + \lambda \frac{\partial g}{\partial y} = 0 \ (2), \quad g = 0 \ (3). \tag{2.32}$$

Since $g(x, y) = y_0(x) - y = 0$, we obtain for (1) and (2)

$$\frac{\partial f}{\partial x} + \lambda \frac{\partial y_0}{\partial x} = 0 \ (1), \quad \frac{\partial f}{\partial y} - \lambda = 0 \ (2). \tag{2.33}$$

Eliminating λ between (1) and (2) obviously amounts to solving the initial equation (2.30).

This method applies in the case of a function $f(\{x_i\})$ of any number of variables x_i, $i = 1, \ldots, n$ related by any number p of constraints $g_k(\{x_i\}) = 0$, $k = 1, \ldots, p$ (with $p < n$).

2.4.4 Boltzmann Factor

It is simpler to work not with the occupation numbers n_i but with the probabilities $p_i = n_i/N$, which can be considered continuous quantities since N is very large.

In terms of these probabilities, the *two* constraints are

$$\sum_i p_i = 1, \quad \sum_i p_i \varepsilon_i = E/N.$$

The function we want to maximize is

$$\ln W = \sum_i n_i \, \ln(N/n_i) = -N \sum_i p_i \, \ln(p_i).$$

We must therefore introduce two Lagrange multipliers, α and β, and the probability law $\{p_i\}$, which maximizes this expression under the above constraints, is the function for which the variation of the quantity

$$\delta \ln W - \alpha \delta N - \beta \delta E \tag{2.34}$$

vanishes. In other words, whatever the δp_i (with $\sum \delta p_i = 0$), the following variation must vanish

$$-\sum_i \delta p_i \, \ln(p_i) - \alpha \sum_i \delta p_i - \beta \sum_i \varepsilon_i \delta p_i. \tag{2.35}$$

We therefore obtain $-\ln p_i - \alpha - \beta \varepsilon_i = 0$; i.e., $p_i = e^{-\alpha - \beta \varepsilon_i}$. The constants α and β are determined by the constraints

$$\sum_i e^{-\alpha - \beta \varepsilon_i} = 1, \quad \sum_i \varepsilon_i e^{-\alpha - \beta \varepsilon_i} = \frac{E}{N}.$$

The condition $\sum_i p_i = 1$ implies

$$e^{-\alpha} = \frac{1}{\sum_i e^{-\beta \varepsilon_i}}. \tag{2.36}$$

Therefore, the equilibrium distribution is characterized by the fact that there exists a number β related to the total energy E by

$$E = N \frac{\sum_i \varepsilon_i e^{-\beta \varepsilon_i}}{\sum_i e^{-\beta \varepsilon_i}}. \tag{2.37}$$

This number β defines the *temperature* T of the assembly by

$$\beta = 1/kT, \tag{2.38}$$

where k is Boltzmann's constant. It is, in particular, a quantity that equalizes when two assemblies are put in thermal contact (which is the first property of temperature).

Therefore, the probability p_i of finding a system in the energy level ε_i at equilibrium is given by Boltzmann's factor

$$p_i = \frac{e^{-\varepsilon_i/kT}}{Z} \qquad (2.39)$$

where the function

$$Z = \sum_i e^{-\varepsilon_i/kT} \qquad (2.40)$$

is called the *partition function* of the system (from the German *Zustandssumme*, sum over states). This function plays an important role in statistical physics; $-k \ln Z$ is the free energy divided by T. The form (2.39) is called the Boltzmann–Gibbs distribution.

2.4.5   Equalization of Temperatures

Consider two assemblies $\mathcal{E}$ and $\mathcal{E}'$, which may be of different natures, formed respectively of N and N' systems S and S'. The energy levels of S are ε_i and those of S' are ε'_j. These two assemblies are in "thermal contact," which means that they can exchange energy but that their interaction is sufficiently weak that it does not change the individual energy levels ε_i and ε'_j of the two systems considered separately.

Furthermore, these two systems are isolated. We denote by E the total energy.

1. The number W of states in a distribution $(\{n_i\}, \{n'_j\})$ of the systems S and S' is

$$W = \frac{N! N'!}{\Pi_i(n_i!) \Pi_j(n'_j!)}.$$

2. There are now three constraints on the distributions $(\{n_i\}, \{n'_j\})$:

$$\sum n_i = N, \quad \sum n'_j = N', \quad \sum n_i \varepsilon_i + \sum n'_j \varepsilon'_j = E.$$

We want to find the distribution that maximizes $\ln W$ under these constraints. We introduce three Lagrange multipliers, α, α', β, and the probabilities $p_i = n_i/N$, $p'_j = n'_j/N$. This leads to finding the zero of

$$\delta \left((W - \alpha N \sum p_i - \alpha' N' \sum p'_j) - \beta (N \sum p_i \varepsilon_i + N' \sum p'_j \varepsilon'_j) \right)$$

or equivalently

$$-N \sum_i \delta p_i \ln(p_i) - N' \sum_j \delta p'_j \ln(p'_j) - \alpha N \sum_i \delta p_i - \alpha' N' \sum_j \delta p'_j$$

$$-\beta \left(N \sum_i \varepsilon_i \delta p_i + N' \sum_j \varepsilon'_j \delta p'_j \right) = 0. \quad (2.41)$$

3. This expression must vanish for all infinitesimal δp_i and $\delta p'_j$. Therefore, one obtains

$$- \ln p_i - \alpha - \beta \varepsilon_i = 0, \quad - \ln p'_i - \alpha' - \beta \varepsilon'_j = 0,$$

or

$$p_i = e^{-\alpha - \beta U_i}, \quad p'_j = e^{-\alpha' - \beta U'_j}. \quad (2.42)$$

We notice that it is the *same* Lagrange multiplier β that appears in both expressions. This is due to the fact that it is the *total energy* that is a given quantity. The constants α, α', and β are fixed by the constraints as above. Therefore, the two temperatures are equal if $\beta = 1/kT$.

2.4.6   The Ideal Gas

Consider a monatomic ideal gas of temperature T. We assume it is confined in a cubic box of side a. The energy levels are therefore

$$\varepsilon_{n,l,m} = \frac{\pi^2 \hbar^2}{2ma^2}(n^2 + l^2 + m^2).$$

We assume that the spacing of these levels is very small compared with the mean thermal energy kT. We go to the continuum limit using the density of states in phase space[6]

$$d^6 N = \frac{d^3 r d^3 p}{2\pi \hbar^3}; \quad \text{i.e.,} \quad d^3 N = \frac{V d^3 p}{(2\pi \hbar)^3},$$

where V is the volume of the cubic container. Inserting this in (2.39), we obtain the probability dP of finding an atom of the gas in an element d^3v around the velocity $\mathbf{v} = \mathbf{p}/m$,

$$dP = \left(\frac{m\beta}{2\pi} \right)^{3/2} exp(-\beta m v^2/2) d^3 v. \quad (2.43)$$

If we identify this with Maxwell's distribution,

---

[6] See, for instance [11], Chap. 4.

$$dP = \left(\frac{m}{2\pi kT}\right)^{3/2} e^{-\frac{mv^2}{2kT}} d^3v,$$

we obtain the fundamental relation

$$\beta = \frac{1}{kT}, \tag{2.44}$$

where T is the absolute temperature.

We now come back to the equalization of the factors β of two assemblies in thermal contact. We see that, since one of these assemblies can be an ideal gas, the energy–absolute-temperature relation holds for any assembly of N identical systems of individual energy levels ε_i,

$$E = N\frac{\sum_i \varepsilon_i e^{-\varepsilon_i/kT}}{\sum_i e^{-\varepsilon_i/kT}} \tag{2.45}$$

(all appropriate care is assumed in the counting of degenerate states and in taking the continuum limit).

Finally, this method allows us to define a *thermostat* by considering the limit where one of the assemblies is much larger than the other. Establishing thermal contact with the second, small assembly does not change the temperature of the first one. We therefore recover the usual treatment of thermodynamics of assemblies in thermal contact with a thermostat at a given temperature.

2.4.7 Boltzmann's Entropy

If we let an assembly evolve freely, it will eventually reach a state of equilibrium. This evolution is not arbitrary: The assembly evolves and reaches its most probable situation where the number of states is maximum, as we have just seen.

It is perhaps the greatest discovery of Boltzmann that the quantity

$$S = k \ln W \tag{2.46}$$

is nothing but the *entropy* of the assembly, which is therefore defined in an *absolute* manner (k is Boltzmann's constant). This provides a measure of the state of *disorder* of the assembly. The greater its disorder, the more stable it is.

This is a fundamental result of great simplicity:

The thermodynamic equilibrium corresponds to a situation that maximizes the entropy for a given set of constraints. In other words, it maximizes the *disorder* given the *constraints*.

This principle applies to a large variety of daily life situations. What is the state of a child's room that optimizes the satisfaction of all the family? Its range of application goes far beyond physics. This notion is one of the founding blocks of economic models.

2.4.8   Heat and Work

The notion of heat, which is very intuitive and has been known since very ancient times, was viewed for a long time as emanating from some fluid that could flow from one body to another. The first principle of thermodynamics tells us that it is a particular form of energy. Statistical thermodynamics allows us to understand this in a very natural manner.

Indeed, consider an assembly at equilibrium whose total energy is $E = N \sum p_i \varepsilon_i$ and whose temperature is T. In any infinitesimal evolution of this assembly through a contact with the outside, two things can happen. One is the variation of the energy levels ε_i if the total volume changes, or if an electric field is applied, etc. Another is the reorganization of the populations of the various energy levels $n_i = N p_i$. The corresponding variation dE of the total energy of the assembly is

$$dE = \sum_i n_i d\varepsilon_i + \sum_i \varepsilon_i N dp_i. \qquad (2.47)$$

The first term is obvious. It corresponds simply to the *work* of the external forces

$$d\tau = \sum_i n_i d\varepsilon_i \qquad (2.48)$$

(we avoid the traditional dW in order to avoid confusion). Under the external action, the energy levels vary, resulting in a variation (2.48) of the total energy of the system.

The second term is less obvious. It comes from the fact that, even if the energy levels ε_i do not change (in the absence of external work), the total energy can be modified by a *rearrangement of the populations* n_i of the levels. This variation of the (internal) energy without any intervention of external forces is what we call "heat." We obtain the statistical definition of heat as

$$dQ = \sum_i \varepsilon_i dn_i. \qquad (2.49)$$

In order to relate Boltzmann's entropy Eq. (2.46) and the usual formula of thermodynamics, consider a variation $dS = kd(\ln W)$. We obtain

$$dS = -kN \sum_i dn_i \ln n_i \qquad (2.50)$$

Fig. 2.8 Simple element of an electrical network with one bifurcation

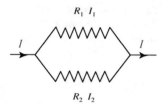

(of course, $\sum_i dn_i = 0$). Suppose the evolution is sufficiently slow that at any time thermodynamical equilibrium is achieved. (The temperature may evolve during the process). This is called a reversible transformation in macroscopic thermodynamics. If this is the case, the n_i are proportional to $\exp(-\varepsilon_i/kT)$, which yields

$$dS_{rev.} = \frac{1}{T}\sum_i \varepsilon_i dn_i. \tag{2.51}$$

If we go back to the definition of heat (2.49), we obtain

$$dS_{rev.} = \left(\frac{dQ}{T}\right)_{rev.}, \tag{2.52}$$

which is exactly the definition of entropy given by Clausius.

We note the following aspects of entropy:

1. Classically, entropy is only defined for systems at equilibrium. The generalization to systems out of equilibrium is completely natural in statistical thermodynamics.
2. The statistical definition expresses the entropy in an *absolute* way. It is not limited to entropy variations.

2.5 Exercises

2.1. Kirchhoff's Laws

We want to determine the relative intensities I_1 and I_2 of the electric current in the two legs of the simple electric circuit shown in Fig. 2.8, whose resistances are R_1 and R_2. The incoming current has an intensity I. The well-known result is easily obtained with the Ohm–Kirchhoff laws.

Show that it is simpler to assume that the loss of energy by Joule heating is minimal.

Fig. 2.9 Soap bubble between two symmetric circles

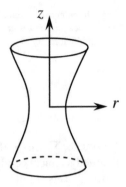

2.2. Shape of a Massive String

Consider a massive string of constant linear mass density μ and length L whose endpoints are fixed at A $(x = 0, z = z_0)$ and B $(x = a, z = z_1)$. The string lies in the vertical plane (x, z), and it is in the gravitational field, oriented along the vertical z axis. Determine the shape of the string at equilibrium. (Of course, we assume that $(z_1 - z_0)^2 + a^2 \leq L^2$.)

2.3. Soap Bubbles

The potential energy of a soap bubble of total area A is $V = \sigma A$, where σ is the surface tension constant of the soap. We consider a soap bubble between two circles of the same axis and same radius R, as in Fig. 2.9. The z axis is the common axis perpendicular to the two circles, which are centered at $z = -h$ and $z = h$, respectively. Find the surface of minimum area attached to the two circles, which are separated by a distance $d = 2h$.

2.4. Conserved Quantities

In the calculation of bent rays (Sect. 2.2.2) or the massive string above, show that the quantity $\Gamma(z) = r/\sqrt{1 + \dot{r}(z)^2}$ is constant along the path (or string). From that observation, deduce the solution of the problem.

2.5. Lagrange Multipliers

Reconsider the massive string exercise above using Lagrange multipliers in order to express the constraints; i.e., the length of the string and the positions of the endpoints.

Fig. 2.10 Definition of
coordinates

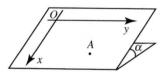

2.6 Problem. Win a Downhill

A skier slides down a snowy plane slope. The plane makes an angle α with respect
to the horizontal direction. The skier is in the vertical field of gravity, of acceleration
g. The skier starts with a zero velocity from some point O and wants to reach a given
point A, downhill, in the shortest time. What is the optimal trajectory?

We choose in the plane a reference frame with origin at O, with horizontal axis
Oy, and whose x axis is along the line of greatest slope, as shown in the Fig. 2.10.
We choose the origin of the potential energy at point O so that the initial energy E
of the skier is zero.

We neglect friction of air and the track, as well as the efforts of the skier to maintain
his trajectory. Therefore, the total energy of the skier is a constant of the motion.

1. Check that with this definition of the variable x, the potential energy of the skier
   at point (x, y) is $V = -mgx \sin \alpha$.
2. Write the expression of the skier's total energy at a given time. We denote $\dot{x} \equiv$
   dx/dt, $\dot{y} \equiv dy/dt$. What is the relation between the potential energy and the
   kinetic energy owing to energy conservation?
3. Use the previous expression to express the square of the time interval dt between
   two positions, (x, y) and $(x + dx, y + dy)$, of the skier, in terms of dx^2, dy^2, x,
   y, g, and α.
4. Calculate the time it takes to go from O to A if the skier follows a trajectory
   defined by a function $y(x)$ (note $y' \equiv dy/dx$).
5. What is the equation of the optimal trajectory?
6. Show that along the optimal trajectory the quantity $C = y'/\sqrt{x(1 + (y')^2)}$ is a
   constant. Deduce from this that along the trajectory the quantity $f(t) = \dot{y}/x$ is a
   constant K, and express its value in terms of C, g, and α.
7. Check that the parametric form $x(\theta) = (1 - \cos 2\theta)/(2C^2)$ and $y(\theta) = (2\theta -$
   $\sin 2\theta)/(2C^2)$ is a solution. Use the result of the previous question to calculate
   the function $\theta(t)$.
8. What kind of curve is it? Draw the trajectory qualitatively in the case $y'(A) \gg 1$.
9. Explain the result physically. (It is not necessary to do all previous calculations
   in order to answer this question.)

Chapter 3
The Analytical Mechanics of Lagrange

In the beginning there was the Action.
Johann Wolfgang Goethe

The fundamental concepts and principles of mechanics, or dynamics, were established in the 17th century. Copernicus (1473–1543) gave the notion of reference system in 1543, and Galileo (1564–1642) stated the principle of inertia in 1638 in his important work *Discorsi e dimostrazioni mathematiche intorno a due nove scienze alla meccanica ed i movimenti locali*.[1] A particle on which no force is exerted has a constant velocity. Linear uniform motion is a *state* relative to the observer, and not a process. It is the variation of the velocity that is a process resulting from an external action.

Many scientists participated in this evolution: Tycho Brahe (1546–1601), Kepler (1571–1630), Father Mersenne, Roberval, Huygens (1629–1695), Varignon etc.

The coronation came with the synthesis of Newton (1643–1727), in 1687, the *Philosophiae Naturalis Principia Mathematica*. Newton laid down his 4 laws: the principle of inertia, the law of composition forces, the proportionality of acceleration and force and the principle of action and reaction. In addition he formulated the universal law of gravitational attraction which enabled him to explain Kepler's laws and the motion of celestial bodies. This celestial motion, completely embedded in the notion of *time*, haunted people since they looked at the sky. Humans now knew how to predict the state of the sky with an incredible accuracy!

But this is by no means the end of the story. Following the Newtonian synthesis, an amazing adventure happened in the 18th and 19th centuries. This started with d'Alembert, Maupertuis, and the Bernoulli brothers (in particular Daniel (1700–

---

[1] Discursive reasoning and mathematical proofs concerning two new sciences of mechanics and local movements.

© The Author(s), under exclusive license to Springer Nature Switzerland AG 2023 37
J.-L. Basdevant, *Variational Principles in Physics*,
https://doi.org/10.1007/978-3-031-21692-3_3

1782)), and was followed by Euler, Lagrange, and later on by Hamilton (1805–1865). The true structure of mechanics was discovered to be a *geometric* structure. A large category of mechanical problems could be reduced to purely geometrical problems.

D'Alembert (1717–1783), who was the first to understand the concept of mass through the notion of linear momentum and its conservation, attacked the abstract concept of force introduced by Newton. For d'Alembert, the only observable phenomenon is motion, whereas the "cause of motion" is an abstraction; hence the idea of studying not a particular trajectory of the theory but the global set of motions that it predicts (this is a very modern conception of forces or interactions).

The crowning achievement of these ideas came with Lagrange in 1788, one century after the *Principia*. Lagrange published, in his *Analytical Mechanics* (*Méchanique Analitique*), a new formulation of mechanics where the global and geometric structure of the theory was emphasized.[2] Lagrange wrote (nevertheless) at that time,

> One will not find Figures in this work. The methods I present do not require any construction or geometric arguments, but only algebraic operations, subject to a uniform and continuous methodology. Those who appreciate Analysis will discover with pleasure how Mechanics becomes a new branch part of it, and they will be grateful to me for extending its realm.

This explains the name "Analytical Mechanics."

Section 3.1 of this chapter describes the principles of the analytical mechanics of Lagrange. It is based on the *least action principle*. Lagrange proposed a new way of considering mechanical problems. Instead of determining the position $\mathbf{r}(t)$ and velocity $\mathbf{v}(t)$ of a particle at time t, given its initial state $\{\mathbf{r}(0), \mathbf{v}(0)\}$, Lagrange wants to determine the *actual trajectory* followed by the particle in order to start at $\mathbf{r}_1$ at time t_1 and to arrive at $\mathbf{r}_2$ at time t_2. This is exactly the same approach that Fermat used for light rays. We will write the Lagrange–Euler equations that elicit the geometrical aspect of the mechanical problem.

Section 3.2 deals with invariance properties of physical phenomena and the resulting conservation laws. Invariance laws are fundamental in the sense that they are what we know a priori about the physics of a problem. We will see that energy conservation is associated with homogeneity of time, momentum conservation with homogeneity of space, and angular momentum conservation with the isotropy of space. In the discussion, we will introduce the notion of Lagrange *conjugate momenta* or *generalized momenta*, which will play a central role in all that follows.

Section 3.3 is devoted to velocity-dependent forces that are not gradients of potentials. We shall say a few words on dissipative systems, but our main point of interest will be the fundamental Lorentz force. The magnetic part of the Lorentz force exerted on a charged particle does not work. We shall see that in that case the generalized momentum does not coincide with the linear momentum (i.e., the product of mass times velocity). This fact, which is intimately related to gauge invariance, has considerable consequences in quantum mechanics and, more generally, in all modern theories of fundamental interactions.

---

[2] There are many books on analytic mechanics, or dynamics. One can refer to the classics of Landau and Lifshitz, [1] and [2], and the book *Classical Mechanics* [3] by Herbert Goldstein, which is clear and complete.

Finally, in Sect. 3.4, we shall extend such considerations to the case of a relativistic particle. We shall restrict ourselves to the case of a massive particle that is free or placed in an electromagnetic field. The basic assumption will be *Lorentz invariance*. In order for the least action principle to have any physical meaning, it must determine the motion of the particle independently of the state of motion of the observer. This will allow us to construct the Lagrangian of a relativistic particle. We will see how the energy and momentum of a free particle are related to its mass and velocity. Two points must be emphasized. First, the Lagrangian formalism allows us to *prove* that the set $\{E/c, \mathbf{p}\}$ is a four-vector of space-time, whereas we have no idea a priori of the values of energy and momentum in terms of the velocity. The second point is that these properties come from the assumption of relativistic invariance of physical laws.

3.1 Lagrangian Formalism and Least Action

In his *Analytic Mechanics*, Lagrange proposes a new way of considering mechanical problems. Instead of determining the position $\mathbf{r}(t)$ and velocity $\mathbf{v}(t)$ of a particle at time t, given its initial state $\{\mathbf{r}(0), \mathbf{v}(0)\}$, Lagrange asks the following question. What is the *actual trajectory* followed by the particle if it starts from $\mathbf{r}_1$ at time t_1 and it arrives at $\mathbf{r}_2$ at time t_2?

3.1.1 Least Action Principle

In order to make things simple, let us consider first the case of only one space dimension. Among the infinite class of possible trajectories (see Fig. 3.1), what is the law that determines the physical one? Lagrange knows that the answer to this question lies in the "principle of natural economy" of Fermat, further developed by Maupertuis, as we said in Chap. 2.

The variational principle we present here is not the original one used by Lagrange; it was formulated by Hamilton in 1834 and is simpler in this discussion. In order not to complicate things, we reverse chronology.

One assumes the following:

1. Any mechanical system is characterized by a Lagrange function, or *Lagrangian* $\mathcal{L}(x, \dot{x}, t)$, which depends on the position x, on its time derivative $\dot{x} = dx/dt$, and possibly on time. The quantities x and $\dot{x}$ are called the state variables of the particle. For a particle in a potential $V(x, t)$, we have for instance

$$\mathcal{L} = \frac{1}{2}m\dot{x}^2 - V(x, t). \tag{3.1}$$

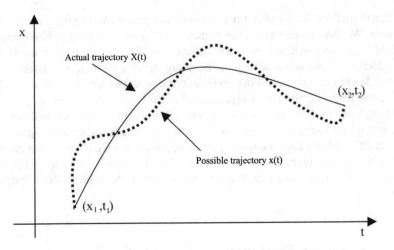

Fig. 3.1 Examples of trajectories starting from x_1 at time t_1 and arriving at x_2 at time t_2. Among all such trajectories, the physical trajectory actually followed by the particle renders the action S minimal (extremal)

2. For *any trajectory* $x(t)$, one can define the *action S* by the integral

$$S = \int_{t_1}^{t_2} \mathcal{L}(x, \dot{x}, t)\, dt. \tag{3.2}$$

The *Least Action Principle* states that the physical trajectory $X(t)$ followed by the particle is such that S is *minimum*, or, more generally, has an extremum.

3.1.2   Lagrange–Euler Equations

We call $X(t)$ the physical trajectory, and we proceed as in Sect. 2.2, except that the variable is now the time t. Consider a trajectory $x(t)$ *infinitely close* to $X(t)$, which also starts from x_1 at t_1 and reaches x_2 at t_2,

$$x(t) = X(t) + \delta x(t), \quad \dot{x}(t) = \dot{X}(t) + \delta\dot{x}(t), \quad \delta\dot{x}(t) = \frac{d}{dt}\delta x(t), \tag{3.3}$$

where by assumption

$$\delta x(t_1) = \delta x(t_2) = 0. \tag{3.4}$$

To first order in δx, the variation of S is

$$\delta S = \int_{t_1}^{t_2} \left(\frac{\partial \mathcal{L}}{\partial x}\, \delta x(t) + \frac{\partial \mathcal{L}}{\partial \dot{x}}\, \delta\dot{x}(t) \right) dt. \tag{3.5}$$

We integrate the second term by parts and take into account (3.4), so that the integrated term vanishes. This leads to

$$\delta S = \int_{t_1}^{t_2} \left(\frac{\partial \mathcal{L}}{\partial x} - \frac{d}{dt} \left(\frac{\partial \mathcal{L}}{\partial \dot{x}} \right) \right) \delta x(t) \, dt. \tag{3.6}$$

The least action principle states that δS must vanish whatever the infinitesimal variation $\delta x(t)$. Therefore, the equation of motion (i.e., the equation that determines the physical trajectory), is the *Lagrange–Euler equation*

$$\frac{\partial \mathcal{L}}{\partial x} = \frac{d}{dt} \left(\frac{\partial \mathcal{L}}{\partial \dot{x}} \right). \tag{3.7}$$

It is simple to check in (3.1) that this leads to the usual equation

$$m\ddot{x} = -\frac{\partial V}{\partial x} \equiv f,$$

where f is the force.

Generalization

The generalization to s degrees of freedom $(x_i, \dot{x}_i)$, $i = 1, \ldots, s$ is straightforward. The Lagrangian is a function $\mathcal{L}(\{x_i\}, \{\dot{x}_i\}, t)$ of the variables $\{x_i\}$ and $\{\dot{x}_i\}$, and the equations of motion are given by the set of Lagrange–Euler equations

$$\frac{\partial \mathcal{L}}{\partial x_i} = \frac{d}{dt} \left(\frac{\partial \mathcal{L}}{\partial \dot{x}_i} \right) \qquad i = 1, \ldots, s. \tag{3.8}$$

In the case of N particles in the usual three-dimensional space, we have $s = 3N$ and can use the notation $(x_i^a, \dot{x}_i^a)$, $a = 1, \ldots, N$, $i = 1, 2, 3$, which leads to

$$\frac{\partial \mathcal{L}}{\partial x_i^a} = \frac{d}{dt} \left(\frac{\partial \mathcal{L}}{\partial \dot{x}_i^a} \right) \qquad i = 1, 2, 3; \, a = 1, \ldots, N. \tag{3.9}$$

Remarks

1. **Nonuniqueness of the Lagrangian**
   The Lagrangian of a given system is by no means unique. It is easy to see that if we add a total derivative with respect to time, i.e.,

$$\mathcal{L}' = \mathcal{L} + \frac{d}{dt} f(\{x_i\}, t),$$

   the equations of motion are unchanged.

2. **Form of the Lagrangian**

   It is mainly invariance considerations that dictate the form of the Lagrangian, in particular translation or rotation invariance. We shall come back to this point. The *kinetic* term $mv^2/2$ comes from the principle of inertia, or equivalently from invariance under Galilean transformations. Consider the simple case of a free particle in space.

   (a) Time has no privileged origin, and therefore $\partial\mathcal{L}/\partial t = 0$.
   (b) Space has no privileged origin, and therefore $\partial\mathcal{L}/\partial x_i) = 0$.
   (c) Rotation invariance implies that $\mathcal{L}$ can only depend on the square of the velocity, $\mathcal{L}(v^2)$.
   (d) Galilean invariance entails that a given law is the same in all Galilean frames. In a reference frame of relative infinitesimal velocity ε, $\mathbf{v}' = \mathbf{v} + \varepsilon$, The Lagrangian becomes, to first order in ε,

   $$\mathcal{L}' = \mathcal{L} + 2\mathbf{v} \cdot \varepsilon \frac{\partial\mathcal{L}}{\partial v^2}.$$

   The second term on the right-hand side is a total derivative with respect to time if and only if $\partial\mathcal{L}/\partial v^2 = $ constant. Therefore, the Lagrangian of a free particle is of the form $\mathcal{L} = Kv^2$, where K is a constant. We choose this constant to be $m/2$ since this entails momentum conservation for an isolated system, as we will see. Therefore, for a free particle,

   $$\mathcal{L} = \frac{m}{2}v^2. \tag{3.10}$$

   (e) In a reference frame with constant velocity $\mathbf{V}$ with respect to the previous one, the Lagrangian becomes

   $$\mathcal{L}' = \frac{m}{2}(\mathbf{v} + \mathbf{V})^2 = \mathcal{L} + \frac{d}{dt}\left(m\mathbf{r} \cdot \mathbf{V} + mV^2\frac{t}{2}\right)$$

   and the equations of motion are the same in both reference frames.

   (f) If the particle is in a field of force, the *potential energy* term in (3.1) is merely a definition of the force. We wish to recover Newton's law, and this choice guarantees it for forces that derive from potentials.

3. **Generalization**

   The Lagrangian of a *set* of N particles in a potential $V(\mathbf{r}_1, \ldots, \mathbf{r}_N; t)$ (which includes the mutual interactions of the particles) is

   $$\mathcal{L} = \frac{1}{2}\sum_{i=1}^{N} m_i(\dot{\mathbf{r}}_i)^2 - V(\mathbf{r}_1, \ldots, \mathbf{r}_N; t). \tag{3.11}$$

4. **Change of System of Coordinates**

The Lagrange–Euler equations keep the same form in all systems of coordinates (for instance $(x, y, z) \rightarrow (r, \theta, \varphi)$). This feature is particularly useful in order to perform changes of variables. One calls a system of coordinates $\{q_i\}$ *generalized* coordinates.

3.1.3 Operation of the Optimization Principle

It is remarkable that the laws of mechanics can be derived from a variational principle. The physical trajectory is that for which the action is *optimal*.

This optimization appears as a "compromise" between various causes in "conflict." Indeed, in the absence of forces ($V = $ constant in (3.1)), S is minimum for $\dot{x} = $ constant, the motion is linear and uniform. In the absence of inertia, on the contrary, the particle would go to the maximum of the potential at the initial point and come back at the final point. The presence of the potential can be considered as a property of *space* that curves the trajectory. Inertia and force can be viewed as conflicting effects. The particle follows a path of minimal "length," this length being measured by the action S.

We see here how the mechanical problem can be transformed into a geometric problem. As we shall see later on, the motion of a particle in a *flat* Euclidean space can be transformed into the *free* motion in a *curved space*, where it moves along geodesics. We will come back extensively to this point in Chap. 7. Einstein had this idea in mind in 1908 when he was constructing general relativity. It took him seven years to elaborate the mathematical details of the final theory.

3.2 Invariances and Conservation Laws

Invariance laws of physical phenomena are fundamental. They form the set of *what is known* a priori about a physical problem. They imply corresponding *conservation laws*, which play a crucial role. In more elaborate problems than those we have seen here, they constitute the guiding line in order to construct the Lagrangian of a system (we have seen a simple example above in discussing the form of the free-particle Lagrangian).

A system with s degrees of freedom possesses a priori $2s$ conserved quantities. Indeed, the evolution of the system is completely determined by the knowledge of the $2s$ initial conditions $\{x_i(0), \dot{x}_i(0)\}$. Therefore, there are in principle $2s$ relations between the variables $\{x_i(t), \dot{x}_i(t)\}$, which allow one to calculate $\{x_i(0), \dot{x}_i(0)\}$ at any time. In practice, only a subset of such relations are usable.

3.2.1   Conjugate Momenta and Generalized Momenta

In order to discuss conservation laws, we introduce the notion of Lagrange *conjugate momenta*. The quantity

$$p_i = \frac{\partial \mathcal{L}}{\partial \dot{x}_i} \tag{3.12}$$

is called the *conjugate momentum* of the variable x_i, or its *generalized momentum*. In the simple case (3.11), this coincides with the linear momentum $p_i = m\dot{x}_i$, but this is no longer true in non-Cartesian coordinate systems, or, as we shall see, when the forces depend on the velocity. We remark that, from (3.7), the time evolution of the conjugate momentum p_i is given by

$$\dot{p}_i = \frac{\partial \mathcal{L}}{\partial x_i}, \tag{3.13}$$

which can be considered the generalized form of Newton's law.

3.2.2   Cyclic Variables

In the Lagrangian formalism, one can make any change of variables

$$(x_1, x_2, \ldots, x_N) \to (q_1, q_2, \ldots, q_N),$$

or

$$\mathcal{L}(\{x_i\}, \{\dot{x}_i\}; t) \to \mathcal{L}'(\{q_i\}, \{\dot{q}_i\}; t).$$

In this change of variables, the Lagrange–Euler equations keep the same form. We can define the conjugate momentum p_i of the generalized variable q_i by the relation

$$p_i = \frac{\partial \mathcal{L}'}{\partial \dot{q}_i}. \tag{3.14}$$

This quantity satisfies the same equation as (3.13); i.e., $\dot{p}_i = \partial \mathcal{L}'/\partial q_i$.

A *cyclic variable* is a variable q_i that does not appear explicitly in the Lagrangian $\mathcal{L}'$. This means that

$$\frac{\partial \mathcal{L}'}{\partial q_i} = 0.$$

In that case, the conjugate momentum $p_i = \partial \mathcal{L}'/\partial \dot{q}_i$ is conserved; we have $p_i =$ constant.

It is useful to find cyclic variables owing to the resulting conservation laws.

3.2.3 Energy and Translations in Time

Assume the system is isolated (i.e., $\partial \mathcal{L}/\partial t = 0$). Another way to describe this assumption is to say that the system is invariant under *translations in time* or that time is homogeneous.

We evaluate the evolution of $\mathcal{L}(x, \dot{x})$ along the physical trajectory $x(t)$,

$$\frac{d\mathcal{L}}{dt}(x, \dot{x}) = \dot{x}(t)\frac{\partial \mathcal{L}}{\partial x} + \ddot{x}(t)\frac{\partial \mathcal{L}}{\partial \dot{x}} = \frac{d}{dt}\left(\dot{x}(t)\frac{\partial \mathcal{L}}{\partial \dot{x}}\right), \tag{3.15}$$

where we have transformed the first term by taking into account the Lagrange equation (3.7). We deduce that

$$\frac{d}{dt}\left(\dot{x}(t)\frac{\partial \mathcal{L}}{\partial \dot{x}} - \mathcal{L}\right) = 0. \tag{3.16}$$

Consequently, for an isolated system, or when there is invariance under translation in time, the quantity

$$E = \dot{x}(t)\frac{\partial \mathcal{L}}{\partial \dot{x}} - \mathcal{L}, \quad \left(\text{resp. } E = \sum_{i=1}^{s} \dot{x}_i(t)\frac{\partial \mathcal{L}}{\partial \dot{x}_i} - \mathcal{L}\right), \tag{3.17}$$

is conserved. It is a *constant of the motion*, the *energy* of the system.

In the general case (3.11), the energy is indeed the sum of the kinetic and potential energies

$$E = \sum_{i=1}^{N} \frac{m_i(\dot{\mathbf{r}}_i)^2}{2} + V(\mathbf{r}_1, \ldots, \mathbf{r}_N; t). \tag{3.18}$$

If we use the Lagrange conjugate momenta, the expression of the energy becomes

$$E = \sum_i p_i \dot{x}_i - \mathcal{L}. \tag{3.19}$$

Examples

Consider the massive string of Chap. 2 and Eq. (A.1). The Lagrangian is (up to factors) $\mathcal{L} \propto z(x)\sqrt{1 + \dot{z}(x)^2}$ (here the variable is x). This Lagrangian does not depend on the variable x. Therefore, the quantity $p\dot{z} - \mathcal{L}$, where p is the conjugate momentum of z, is constant along the curve (it is a "constant of the motion" in the language of the present chapter). One obtains with no difficulty $p = z\dot{z}/\sqrt{1 + \dot{z}(x)^2}$ and $p\dot{z} - \mathcal{L} = -z/\sqrt{1 + \dot{z}(x)^2} = -c$, where c is a constant.

We deduce from this that

$$z(x) = c\sqrt{1 + \dot{z}(x)^2}.$$

If, by the definition of $\phi(x)$, we set $\dot{z}(x) = \sinh(\phi(x))$, we obtain, by inserting this into the equation,

$$z(x) = c\cosh(\phi(x)); \quad \text{i.e.,} \quad \dot{z} = c\dot{\phi}(x)\sinh(\phi(x)).$$

We conclude that $c\dot{\phi}(x) = 1$, and the solution given in (2.17) follows: $z(x) = c\cosh((x - x_0)/c)$. Using this conserved quantity simplifies the resolution of the problem.

In general, if we consider a Lagrangian of the form

$$\mathcal{L} = f(z(x))\sqrt{1 + \dot{z}(x)^2}, \tag{3.20}$$

the conjugate momentum of z is

$$p = \frac{f(z)\dot{z}}{\sqrt{1 + \dot{z}(x)^2}}. \tag{3.21}$$

Since the Lagrangian does not depend explicitly on variable x, the quantity

$$A = p\dot{z} - \mathcal{L} = -\frac{f(z)}{\sqrt{1 + \dot{z}(x)^2}}, \tag{3.22}$$

whose value is fixed by the initial conditions, is conserved. We therefore obtain

$$A^2(1 + \dot{z}^2) = f(z)^2; \quad \text{i.e.,} \quad \dot{z} = \sqrt{\left(\frac{f(z)}{A}\right)^2 - 1}. \tag{3.23}$$

The general solution amounts to a simple quadrature

$$A^2(1 + \dot{z}^2) = f(z)^2; \quad \text{i.e.,} \quad \int_{z_0}^{z} \frac{dz}{\sqrt{(\frac{f(z)}{A})^2 - 1}} = x - x_0. \tag{3.24}$$

This is the generalization of the usual method of integration of the equation of motion when there is energy conservation.

3.2.4 Noether Theorem: Symmetries and Conservation Laws

The 1918 theorem, established by Emmy Noether, is of paramount importance in contemporary physics.[3] It expresses that the invariance of physical laws under transformations, called symmetries, leads to the existence of laws of conservation of physical quantities. Consider a family, with a continuous parameter s, of generalized coordinate transformations.

$$q_i(t) \rightarrow Q_i(s, t) \quad s \in \mathcal{R} \quad \text{with} \quad Q_i(0, t) = q_i(t) . \qquad (3.25)$$

This transformation is called a symmetry of the lagrangian $\mathcal{L}$ if

$$\frac{d}{ds} \mathcal{L}(Q_i(s, t), \dot{Q}_i(s, t), t) = 0 \qquad (3.26)$$

The Noether theorem says that to any symmetry of this type, there corresponds a conserved quantity.

The demonstration is as follows. By hypothesis, since $\mathcal{L}$ does not depend on s, we have

$$\frac{d\mathcal{L}}{ds} = \frac{\partial \mathcal{L}}{\partial Q_i} \frac{\partial Q_i}{\partial s} + \frac{\partial \mathcal{L}}{\partial \dot{Q}_i} \frac{\partial \dot{Q}_i}{\partial s} , \qquad (3.27)$$

and, for $s = 0$,

$$0 = \frac{d\mathcal{L}}{ds}\Big|_{s=0} = \frac{\partial \mathcal{L}}{\partial q_i} \frac{\partial Q_i}{\partial s}\Big|_{s=0} + \frac{\partial \mathcal{L}}{\partial \dot{q}_i} \frac{\partial \dot{Q}_i}{\partial s}\Big|_{s=0} . \qquad (3.28)$$

Making use of Lagrange–Euler equations, this leads to

$$\frac{d}{dt}\left(\frac{\partial \mathcal{L}}{\partial \dot{q}_i}\right) \frac{\partial Q_i}{\partial s}\Big|_{s=0} + \frac{\partial \mathcal{L}}{\partial \dot{q}_i} \frac{\partial \dot{Q}_i}{\partial s}\Big|_{s=0} = \frac{d}{dt}\left(\frac{\partial \mathcal{L}}{\partial \dot{q}_i} \frac{\partial Q_i}{\partial s}\Big|_{s=0}\right) = 0 . \qquad (3.29)$$

In other words, the quantity

$$\sum_i \left(\frac{\partial \mathcal{L}}{\partial \dot{q}_i}\right)\left(\frac{\partial Q_i}{\partial s}\Big|_{s=0}\right) \qquad (3.30)$$

is a time independent constant.

---

[3] A complete account can be found in Yvette Kosmann-Schwartzbachs book [10].

3.2.5 Momentum and Translations in Space

Suppose the problem is invariant under translations in space. This is the case for a free particle, and it is also the case for a system of particles whose interactions depend only on the relative coordinates: $V(\{\mathbf{r}_i - \mathbf{r}_j\})$.

In this case, for any infinitesimal transformation $\mathbf{r}_i \to \mathbf{r}_i + \varepsilon$, the Lagrangian is invariant:

$$\delta\mathcal{L} = \sum_i \frac{\partial \mathcal{L}}{\partial \mathbf{r}_i} \cdot \varepsilon = 0 \quad \forall''; \quad \text{i.e.,} \quad \sum_i \frac{\partial \mathcal{L}}{\partial \mathbf{r}_i} = 0. \tag{3.31}$$

For convenience, we use the notation

$$\frac{\partial \Phi}{\partial \mathbf{r}_i} \equiv \nabla_i \Phi, \tag{3.32}$$

where the gradient is taken with respect to the vector variable $\mathbf{r}_i$.

If the Lagrangian of the system is of the form (3.11), this relation is simply the principle of action and reaction of Newton. Indeed, if we consider a system of two particles interacting via a potential $V(\mathbf{r}_1 - \mathbf{r}_2)$, we obtain

$$\mathbf{f}_1 = -\nabla_1 V = +\nabla_2 V = -\mathbf{f}_2. \tag{3.33}$$

However, there is another interpretation of the result (3.31). Using the definitions (3.12) and (3.13) of the momenta and their time derivatives, this relation can be written as

$$\frac{d}{dt} \sum_{i=1}^{N} \mathbf{p}_i \equiv \frac{d}{dt} \mathbf{P} = 0, \tag{3.34}$$

where $\mathbf{P}$ is the total momentum $\mathbf{P} = \sum_{i=1}^{N} \mathbf{p}_i$.

Translation invariance in space implies conservation of the *total momentum* of a system of particles.

3.2.6 Angular Momentum and Rotations

Consider rotations. An infinitesimal rotation of an angle $\delta\phi$ around an axis along the unitary vector $\hat{u}$ transforms positions and velocities as

$$\mathbf{r}_i \to \mathbf{r}_i + \delta\phi\, \hat{u} \times \mathbf{r}_i, \quad \dot{\mathbf{r}}_i \to \dot{\mathbf{r}}_i + \delta\phi\, \hat{u} \times \dot{\mathbf{r}}_i.$$

In this transformation, the variation of the Lagrangian is

$$\delta \mathcal{L} = \sum_i \left(\frac{\partial \mathcal{L}}{\partial \mathbf{r}_i} \cdot (\delta \phi \, \hat{u} \times \mathbf{r}_i) + \frac{\partial \mathcal{L}}{\partial \dot{\mathbf{r}}_i} \cdot (\phi \, \hat{u} \times \dot{\mathbf{r}}_i) \right). \tag{3.35}$$

If there is rotation invariance, $\delta \mathcal{L} = 0$ for all $\delta \phi \, \hat{u}$. Coming back to the definition of conjugate momenta and their derivatives, we obtain

$$\sum_i (\mathbf{r}_i \times \dot{\mathbf{p}}_i + \dot{\mathbf{r}}_i \times \mathbf{p}_i) = 0.$$

In other words,

$$\frac{d}{dt} \left(\sum_i (\mathbf{r}_i \times \mathbf{p}_i) \right) \equiv \frac{d}{dt} \left(\sum_i \mathbf{L}_i \right) = \frac{d}{dt} \mathbf{L} = 0, \tag{3.36}$$

where the angular momentum $\mathbf{L}_i$ of each particle and the total angular momentum $\mathbf{L}$ are defined by

$$\mathbf{L}_i = \mathbf{r}_i \times \mathbf{p}_i, \qquad \mathbf{L} = \sum_i \mathbf{L}_i. \tag{3.37}$$

To rotation invariance there corresponds the conservation of the *total angular momentum*.

3.2.7 Dynamical Symmetries

A problem may have symmetries of dynamical origin, which can be more or less hidden. We will examine in Chap. 4 some of the many symmetries of the harmonic oscillator.

The Kepler problem $V(r) = -g^2/r$ and $\mathcal{L} = mv^2/r + g^2/r$ has a well-known symmetry that comes from the conservation of the Lenz vector. This vector is

$$\mathbf{A} = \frac{\mathbf{p} \times \mathbf{L}}{m} - g^2 \frac{\mathbf{r}}{r}, \tag{3.38}$$

where $\mathbf{p}$ is the momentum and $\mathbf{L} = \mathbf{r} \times \mathbf{p}$ the angular momentum of the particle.

In Kepler's problem, we must determine six quantities as a function of time $\mathbf{r}(t), \dot{\mathbf{r}}(t)$. Conservation of angular momentum and energy fixes four of them. The conservation of the Lenz vector, which is perpendicular to the angular momentum and therefore lies in the plane of the trajectory, fixes the two others. Therefore, the solution of the problem does not necessitate any quadrature. One consequence is that in the case of bound states, the trajectory is closed, which is exceptional: Only the harmonic potential ($\propto r^2$) and the Newtonian potential ($\propto 1/r$) lead to this property.

To this conservation law corresponds an invariance law of the lagrangian. Going back to this invariance is a nontrivial problem of group theory. A complete account is given by Bander and Itzykson[4] and in the book by Alain Guichardet.[5] In fact, in Keplers problem there are two types of angular momenta, **L** and **A**, which are related in a subtle manner. The result is a four-dimensional rotational symmetry of the problem, in mathematics an $O(4)$ symmetry.

3.3 Velocity-Dependent Forces

3.3.1 Dissipative Systems

One can convince oneself that the formalisms of Lagrange and Newton coincide in the case of conservative forces, which derive from potentials. However, the Lagrangian formalism does not easily accommodate dissipative forces that depend on the velocity, such as friction. Dissipative forces belong to the mechanics of continuous media, and we are not much concerned with that here.[6]

We can nevertheless give, as a concrete example, a Lagrangian method that can deal with simple dissipative systems by a trick. Consider a system that loses energy by friction, Joule heating, or any other process. The trick consists in coupling the system appropriately with a fictitious *mirror system* that formally *absorbs* the energy in such a way that the total energy of the two systems remains constant. Naturally, one only attributes a physical meaning to quantities or results that possess one.

Consider, for definiteness, a damped harmonic oscillator in one dimension, of coordinate x, whose equation of motion is

$$m\ddot{x} + R\dot{x} + kx = 0. \tag{3.39}$$

In order to obtain this result in the Lagrangian formalism, we introduce a "mirror" oscillator of coordinate x^* and the formal Lagrangian for the set of the two coupled systems

$$\mathcal{L} = m(\dot{x}\dot{x}^*) - \frac{1}{2}R(x^*\dot{x} - x\dot{x}^*) - kxx^*. \tag{3.40}$$

The conjugate momenta are

$$p = m\dot{x}^* - Rx^*/2 \quad \text{and} \quad p^* = m\dot{x} + Rx/2. \tag{3.41}$$

---

[4] M. Bander and C. Itzykson, Rev. Mod. Phys. 38, 330 (1966).

[5] Alain Guichardet, *Le Probleme de Kepler Histoire & Theorie*, Editions de l'Ecole Polytechnique (2012).

[6] For a general treatment of dissipative forces, we refer to Chap. 3, Sect. 2, of Morse and Feshbach [12].

They have nothing to do with the linear momentum of the damped oscillator (3.39).

Applying the Lagrange–Euler equations to the two variables $\{x, x^*\}$, one obtains the two equations

$$m\ddot{x} + R\dot{x} + kx = 0 \quad \text{and} \quad m\ddot{x}^* - R\dot{x}^* + kx^* = 0. \tag{3.42}$$

The second equation represents an oscillator that "absorbs" the energy lost by the damping of the first one.

The energy of the set of two systems

$$E = p\dot{x} + p^*\dot{x}^* - \mathcal{L} = \dot{x}\dot{x}^* + kxx^* \tag{3.43}$$

is a constant of the motion. The amplitude of the variable x^* increases as fast as that of x decreases.

3.3.2 Lorentz Force

On the other hand, fundamental interactions provide us with a velocity-dependent conservative force, the Lorentz (1853–1928) force, whose magnetic part does not work.

Linear Term in the Velocity

In order to prepare the argument, consider, in three-dimensional space, the case of a Lagrangian whose potential part is linear in the velocity

$$\mathcal{L} = \frac{1}{2}m\dot{r}^2 + \dot{\mathbf{r}} \cdot \mathbf{A}(\mathbf{r}, t), \tag{3.44}$$

where $\mathbf{A}(\mathbf{r}, t)$ is a given vector field.

The Lagrange equations give, for the x component for instance,

$$\dot{\mathbf{r}} \cdot \frac{\partial \mathbf{A}(\mathbf{r}, t)}{\partial x} = m\ddot{x} + \frac{d}{dt}A_x(\mathbf{r}, t). \tag{3.45}$$

Taking into account

$$\frac{d}{dt}A_x(\mathbf{r}, t) = \dot{x}\frac{\partial A_x(\mathbf{r}, t)}{\partial x} + \dot{y}\frac{\partial A_x(\mathbf{r}, t)}{\partial y} + \dot{z}\frac{\partial A_x(\mathbf{r}, t)}{\partial z} + \frac{\partial A_x(\mathbf{r}, t)}{\partial t}, \tag{3.46}$$

we obtain

$$m\ddot{x} = \dot{y}\left(\frac{\partial A_y(\mathbf{r}, t)}{\partial x} - \frac{\partial A_x(\mathbf{r}, t)}{\partial y}\right) - \dot{z}\left(\frac{\partial A_x(\mathbf{r}, t)}{\partial z} - \frac{\partial A_z(\mathbf{r}, t)}{\partial x}\right) - \frac{\partial A_x(\mathbf{r}, t)}{\partial t}.$$

(3.47)

Therefore, if we introduce the vector field

$$\mathbf{B}(\mathbf{r}, t) = \nabla \times \mathbf{A}(\mathbf{r}, t),$$

(3.48)

we obtain the vector expression

$$m\ddot{\mathbf{r}} = \dot{\mathbf{r}} \times \mathbf{B}(\mathbf{r}, t) - \frac{\partial \mathbf{A}(\mathbf{r}, t)}{\partial t},$$

(3.49)

whose form is of obvious interest.

Maxwell's Equations and the Lorentz Force

Classically, a particle of charge q placed in an electromagnetic field undergoes the Lorentz force

$$\mathbf{f} = q\ (\mathbf{E} + \mathbf{v} \times \mathbf{B}).$$

This force is velocity dependent and does not derive from a potential. The magnetic part $q\mathbf{v} \times \mathbf{B}$ does not work. Let Φ be the electric potential. The potential energy boils down to its electric part $V = q\Phi$, and the total energy is $E = m\mathbf{v}^2/2 + q\Phi$.

The Lagrangian cannot be $\mathcal{L} = m\mathbf{v}^2/2 - q\Phi$ since one would lose track of the magnetic field.

The result (3.49) shows how a linear dependence of the Lagrangian on the velocity can solve this problem, owing to the properties of the electromagnetic field.

Maxwell's homogeneous equations

$$\nabla \cdot \mathbf{B} = 0, \qquad \nabla \times \mathbf{E} = -\frac{\partial \mathbf{B}}{\partial t},$$

(3.50)

allow us to express the fields $\mathbf{E}$ and $\mathbf{B}$ in terms of the scalar and the vector potentials Φ and $\mathbf{A}$,

$$\mathbf{B} = \nabla \times \mathbf{A}, \qquad \mathbf{E} = -\nabla\Phi - \frac{\partial \mathbf{A}}{\partial t}.$$

(3.51)

Consider a particle of mass m and charge q placed in this electromagnetic field. We note as usual $\mathbf{r}$ and $\dot{\mathbf{r}} = \mathbf{v}$, the position and velocity of the particle.

A possible Lagrangian for this particle is expressed in terms of the *potentials* $\mathbf{A}$ and Φ,

$$\mathcal{L} = \frac{1}{2}m\dot{\mathbf{r}}^2 + q\,\dot{\mathbf{r}} \cdot \mathbf{A}(\mathbf{r}, t) - q\,\Phi(\mathbf{r}, t).$$

(3.52)

As in (3.49), one can verify by using the Lagrange equations and

$$\frac{d}{dt}\mathbf{A}(\mathbf{r}, t) = \frac{\partial \mathbf{A}}{\partial t} + \dot{x}\frac{\partial \mathbf{A}}{\partial x} + \dot{y}\frac{\partial \mathbf{A}}{\partial y} + \dot{z}\frac{\partial \mathbf{A}}{\partial z}$$

that we obtain the required equation of motion $m\ddot{\mathbf{r}} = q(\mathbf{E} + \dot{\mathbf{r}} \times \mathbf{B})$.

3.3.3 Gauge Invariance

One thing may, however, seem surprising. We have expressed the Lagrangian in terms of the potentials Φ and $\mathbf{A}$. However, these are not unique. The fields $\mathbf{E}$ and $\mathbf{B}$ are invariant under *gauge transformations*,

$$\mathbf{A} \to \mathbf{A}' = \mathbf{A} + \nabla\chi(\mathbf{r}, t), \qquad \Phi \to \Phi' = \Phi - \frac{\partial \chi}{\partial t}, \qquad (3.53)$$

where $\chi(\mathbf{r}, t)$ is an arbitrary function.

If we insert this transformation in (3.52), we obtain

$$\mathcal{L}' = \mathcal{L} + q\left(\dot{\mathbf{r}} \cdot \nabla\chi(\mathbf{r}, t) + \frac{\partial \chi}{\partial t}\right). \qquad (3.54)$$

The difference is a total time derivative

$$\mathcal{L}' = \mathcal{L} + q\frac{d}{dt}\chi(\mathbf{r}, t). \qquad (3.55)$$

Therefore, a gauge transformation does not affect the physics of the problem. This is of course obvious in the equations of motion. It becomes less obvious when one transposes the result in quantum mechanics.[7] Gauge invariance is a dynamical symmetry that one can visualize as defining field theories. This is the starting point of modern theories of fundamental interactions.

3.3.4 Momentum

Consider now the *conjugate momentum* $\mathbf{p}$. From the definition (3.12), we obtain

$$\mathbf{p} = m\dot{\mathbf{r}} + q\mathbf{A}(\mathbf{r}, t). \qquad (3.56)$$

---

[7] See, for instance A. Tonomura et al, *Evidence for Aharonov-Bohm effect with magnetic field completely shielded from electron wave* Phys. Rev. Lett 56, 792 (1986).

In other words, in a magnetic field, the momentum **p** *does not coincide* with the linear momentum $m\dot{\mathbf{r}}$!

Similarly, the angular momentum $\mathbf{L} = \mathbf{r} \times \mathbf{p}$ does not coincide with $\mathbf{r} \times m\mathbf{v}$.

Finally, note that the relation (3.56) and our result (3.52) simply combine in the expression of the energy (3.19) of a charged particle placed in the potentials Φ and **A**. The corresponding *Hamitonian H*, which we will talk about in the following chapters, is:

$$H = \frac{1}{2m}(\mathbf{p} - q\mathbf{A})^2 + q\Phi \ . \tag{3.57}$$

3.4 Lagrangian of a Relativistic Particle

The Lagrangian formalism has the fundamental physical property (which could not be guessed, neither by Lagrange nor by Maxwell) that it can be extended directly to the case of a relativistic particle because it is formulated directly in space and time. Therefore it can be written in the four-dimensional space-time of Minkowski, and can become the relativistic theory of a charged particle motion. Well see this here on the case of a massive particle free, or placed in an electromagnetic field.

3.4.1 Lorentz Transformation

We want to express Lorentz invariance.

The principle of least action must determine the motion of the particle, whatever the relative state of motion of the observer.

A Lorentz transformation between the between the coordinates (x, y, z, t) and (x', y', z', t') of the same event, measured by two observers in uniform rectilinear motion of velocity v along the x axis relative to each other, is of the form:

$$\begin{cases} x' = \gamma(x - vt) & \text{with} \ \ \gamma = 1/\sqrt{1 - (v^2/c^2)} \\ y' = y \\ z' = z \\ t' = \gamma(t - (vx)/c^2) \ . \end{cases} \tag{3.58}$$

We recall that Minkowskis space-time is a four-dimensional vector space with a (pseudo-)lorentzian scalar product of signature $(+, -, -, -)$.[8] An orthogonal basis consisting of four vectors: $\mathbf{e}_i$, $i = 0, 1, 2, 3$ which have the orthogonality relations

$$\mathbf{e}_i.\mathbf{e}_j = \eta_{ij} \quad \text{with} \quad \eta_{00} = +1, \quad \eta_{ij} = -\delta_{ij}, \quad \eta_{ij} = 0 \quad \text{if } i \neq j \ .$$

---

[8] This signature convention will be modified in Chaps. 7 and 8.

We note $x^i = (x^0, x^1, x^2, x^3) = (ct, x, y, z) \equiv (ct, \mathbf{r})$ the components (t, x, y, z) of a *four-vector* of space-time, placing time first (multiplied by c for homogeneity), the spatial part is a usual vector of $\mathcal{R}^3$. In general a four-vector is a vector of Minkowski space, on which the changes of reference system are made by Lorentz transformations. The scalar product of two fourvectors a^i and b^i is, following the Einstein convention of summation on the repeated up and down indices,

$$\eta_{ij} a^i b^j = a^0 b^0 - \mathbf{a}.\mathbf{b} \quad . \tag{3.59}$$

Note, in the above Lorentz transformation, that we have between the duration and the distance of the two observed events the relation:

$$c^2 t^2 - \mathbf{r}^2 = c^2 t'^2 - \mathbf{r}'^2 \quad . \tag{3.60}$$

This quantity is a relativistic invariant, common to both observations, it is, up to the factor c, the square of the *proper time* τ observed by an object in motion between the beginning and the end of its space-time trajectory, and measured by the two observers:

$$\tau^2 = t^2 - \mathbf{r}^2/c^2 = t'^2 - \mathbf{r}'^2/c^2 \quad . \tag{3.61}$$

The argument is based on Lorentz invariance. The least action principle must yields the same equation of motion whatever the relative state of free motion of the observer. We proceed as in Sect. 3.1. We want to determine the path followed to go from $A(\mathbf{r}_1, t_1)$ to $B(\mathbf{r}_2, t_2)$ by minimizing the action

$$S = \int_{t_1}^{t_2} \mathcal{L}(\mathbf{r}, \dot{\mathbf{r}}) \, dt. \tag{3.62}$$

3.4.2 Free Particle

Consider first a free particle of mass m. We know the result: The motion is linear and uniform.

Among all possible paths, the free motion corresponds to *the largest proper time*.

Let dt be the time interval measured by an observer with a relative velocity v with respect to the particle. The *proper time* of the particle is $d\tau = dt \sqrt{1 - v^2/c^2}$. Therefore, free motion maximizes the quantity

$$\tau = \int_{t_1}^{t_2} \sqrt{\left(1 - \frac{v^2}{c^2}\right)} \, dt \tag{3.63}$$

which is, by construction, Lorentz invariant.

In order to recover a minimization principle and to obtain the non-relativistic limit, which we already know, we choose the Lagrangian

$$\mathcal{L} = -mc^2 \sqrt{1 - \frac{v^2}{c^2}}. \tag{3.64}$$

The action is

$$S = -mc^2 \int_{t_1}^{t_2} \sqrt{1 - \frac{v^2}{c^2}} \, dt. \tag{3.65}$$

This action is Lorentz invariant, whereas the Lagrangian (3.64) is not. This comes from the fact that in the present approach, time, over which we integrate, plays a special role. One can get rid of this problem, but we shall not do it here.

We remark that in the limit of small velocities, we recover the non-relativistic Lagrangian up to a constant: $\mathcal{L} = -mc^2 + mv^2/2$.

3.4.3 Energy and Momentum

We deduce the expression of the energy and momentum by following the same method as in Sect. 3.2. These quantities are of interest since they are conserved if space-time is homogeneous. This holds in any reference system.

The conjugate momentum is

$$\mathbf{p} = \frac{\partial \mathcal{L}}{\partial \mathbf{v}} = \frac{m\mathbf{v}}{\sqrt{1 - v^2/c^2}}. \tag{3.66}$$

The energy is

$$E = \mathbf{p} \cdot \mathbf{v} - \mathcal{L} = \frac{mc^2}{\sqrt{1 - \frac{v^2}{c^2}}} \quad \text{or also} \quad E = \sqrt{p^2c^2 + m^2c^4}. \tag{3.67}$$

We see that the set of four quantities $\{E/c, \mathbf{p}\}$ is a four-vector of space-time. Energy and momentum satisfy the relation

$$(E/c)^2 - p^2 = m^2c^2. \tag{3.68}$$

Two points must be emphasized. First, the Lagrangian formalism allows us to *prove* that the set $\{E/c, \mathbf{p}\}$ is a four-vector of space-time, whereas neither energy nor momentum are defined a priori, and we have only worked with positions and velocities. Second, this property follows, of course, from our starting assumption (3.65) based on the relativistic invariance of physical laws.

The observed velocity of the particle is related to its momentum and to its energy by

$$\mathbf{v} = \frac{\mathbf{p}c^2}{E}.$$

(3.69)

Let us note a simple but amazing result. The duration of an event depends only on the velocity and not on the distance travelled, even if this speed varies (3.63 is then a sum). The lepton μ or *muon*, a heavy electron congener, with a mass 200 times greater than the electron $m_\mu c^2 \sim 105\,\mathrm{MeV}$, has a mean lifetime of 2.2μs. This particle is studied at Fermilab in a high energy storage ring with a diameter of 35 meters around which are positioned superconducting magnets that produce a constant, uniform and very precise magnetic field. The lifetime measured at this energy (Lorentz factor $\gamma \sim 30$) with this magnetic field is perfectly consistent with its value at rest deduced from its circular velocity while the muons undergo a radial acceleration greater than $10^{18}\,g$. In other words, the magnetic field does not alter its variation.

3.4.4 Interaction with an Electromagnetic Field

Consider now a particle of electric charge q and mass m, in an electromagnetic field.

We want to determine the form of the interaction Lagrangian $\mathcal{L}_I$ of the particle and the field. If we insert the sum $(\mathcal{L} + \mathcal{L}_I)$ in (3.62), the Lagrange–Euler equations must give us the equation of motion $\dot{\mathbf{p}} = \mathbf{f}$, where $\mathbf{f}$ is the Lorentz force.

Actually, we already know the answer because the interaction part in (3.52) is a relativistic formula!

One can recover this form using relativistic invariance considerations.

- We want $\mathcal{L}_I dt$ to be Lorentz invariant, as is $\mathcal{L}dt$ above.
- Consider the potential four-vector $A^\mu = (\phi/c, \mathbf{A})$ of the electromagnetic field and the velocity four-vector of the particle $u^\mu = (\gamma c, \gamma \mathbf{v})$. The scalar product

$$u^\mu A_\mu = \gamma(\phi - \mathbf{v} \cdot \mathbf{A})$$

(3.70)

is a Lorentz invariant quantity.
- Therefore, the quantity

$$\mathcal{L}_I = -q\, u^\mu A_\mu / \gamma = q\,(\mathbf{v} \cdot \mathbf{A} - \phi)$$

(3.71)

has the properties wanted. This expression is identical to the interaction term of (3.52). It is called the "minimal interaction" of a charged particle with an electromagnetic field.

We therefore obtain the expression of the relativistic Lagrangian of a particle of mass m and charge q in an electromagnetic field that derives from the potentials ϕ and $\mathbf{A}$,

$$\mathcal{L} = -mc^2 \sqrt{1 - \frac{v^2}{c^2}} + q(\mathbf{v} \cdot \mathbf{A} - \phi). \tag{3.72}$$

The equation of motion follows from the standard procedure.

1. **Conjugate Momentum**

   Let **p** be the momentum in the absence of the field, as defined by (3.66):

$$\mathbf{p} = \frac{m\mathbf{v}}{\sqrt{1 - v^2/c^2}}. \tag{3.73}$$

   The conjugate momentum $\mathbf{P} = \partial\mathcal{L}/\partial\mathbf{v}$ is related to this momentum **p** by

$$\mathbf{P} = \frac{\partial\mathcal{L}}{\partial\mathbf{v}} = \mathbf{p} + q\mathbf{A}. \tag{3.74}$$

2. **Lagrange–Euler Equations**

   The equation of motion follows from the Lagrange–Euler equations.

$$\frac{d}{dt}\left(\frac{\partial\mathcal{L}}{\partial\mathbf{v}}\right) = \frac{\partial\mathcal{L}}{\partial\mathbf{r}}. \tag{3.75}$$

   We have

$$\frac{\partial\mathcal{L}}{\partial\mathbf{r}} = q(\nabla(\mathbf{v} \cdot \mathbf{A}) - \nabla\phi), \tag{3.76}$$

   which yields

$$\frac{d\mathbf{P}}{dt} = \frac{d(\mathbf{p} + q\mathbf{A})}{dt} = q(\nabla(\mathbf{v} \cdot \mathbf{A}) - \nabla\phi). \tag{3.77}$$

3. **Equations of Motion**

   We use the relations

$$\frac{d\mathbf{A}}{dt} = \frac{\partial\mathbf{A}}{\partial t} + \left(\dot{x}\frac{\partial\mathbf{A}}{\partial x} + \dot{y}\frac{\partial\mathbf{A}}{\partial y} + \dot{z}\frac{\partial\mathbf{A}}{\partial z}\right) = \frac{\partial\mathbf{A}}{\partial t} + (\mathbf{v} \cdot \nabla)\mathbf{A}, \tag{3.78}$$

   and

$$\nabla(\mathbf{v} \cdot \mathbf{A}) = (\mathbf{v} \cdot \nabla)\mathbf{A} + \mathbf{v} \times (\nabla \times \mathbf{A}). \tag{3.79}$$

   This leads to the equation of motion

$$\frac{d\mathbf{p}}{dt} = q(\mathbf{E} + \mathbf{v} \times \mathbf{B}), \tag{3.80}$$

   where the momentum **p** and the velocity **v** are related by (3.73).

4. We must take care of the relation (3.73). If we define the kinetic energy $\mathcal{E}_{kin}$ by

$$\mathcal{E}_{kin} = \frac{mc^2}{\sqrt{1 - \frac{v^2}{c^2}}}, \tag{3.81}$$

by taking the derivative of this equation with respect to time, and taking into account the definition (3.73), we obtain

$$\frac{d\mathcal{E}_{kin}}{dt} = \mathbf{v} \cdot \frac{d\mathbf{p}}{dt}. \tag{3.82}$$

Inserting this in equation (3.80) and taking into account that $\mathbf{v} \cdot (\mathbf{v} \times \mathbf{B}) = 0$, we obtain the anticipated result

$$\frac{d\mathcal{E}_{kin}}{dt} = q\mathbf{v} \cdot \mathbf{E}, \tag{3.83}$$

where $\mathbf{E}$ is the electric field. Only the electric field works and modifies the kinetic energy and the norm of the velocity.

3.5 Exercises

3.1. Sliding Pendulum

Consider a pendulum of length l and mass m_2 hanging on a point of mass m_1 that moves horizontally without friction on a rail. We note x the abscissa of m_1 and ϕ the angle with the vertical direction. Write the Lagrangian of this system.

3.2. Properties of the Action

1. We consider a free particle of Lagrangian $\mathcal{L} = m\dot{x}^2/2$. Calculate the action along the physical trajectory in terms of the positions and times of departure (x_1, t_1) and arrival (x_2, t_2).
2. Consider a harmonic oscillator $\mathcal{L} = (m/2)(\dot{x}^2 - \omega^2 x^2)$. Calculate the action, setting $T = t_2 - t_1$.
3. Calculate the action for a constant force $\mathcal{L} = m\dot{x}^2/2 - Fx$.
4. Show that the momentum at the point of arrival x_2 is given by

$$p_2 = \left(\frac{\partial \mathcal{L}}{\partial \dot{x}}\right)_{x=x_2} = \frac{\partial S_{12}}{\partial x_2}.$$

5. Show that the energy $E = p\dot{x} - \mathcal{L}$ at the point of arrival x_2 is given by

$$E_2 = -\frac{\partial S_{12}}{\partial t_2}.$$

3.3. Conjugate Momenta in Spherical Coordinates

We consider a non-relativistic particle of mass m in a central potential $V(r)$, where $r = \sqrt{x^2 + y^2 + z^2}$. We denote the velocity $\mathbf{v} \equiv \dot{\mathbf{r}}$ and v^2 its square.

We study the problem in spherical coordinates (r, θ, ϕ) defined by

$$x = r \sin\theta \cos\phi, \quad y = r \sin\theta \sin\phi, \quad z = r\cos\theta. \tag{3.84}$$

The square of the velocity is

$$v^2 = \dot{r}^2 + r^2\,\dot{\theta}^2 + r^2 \sin^2\theta\,\dot{\phi}^2. \tag{3.85}$$

1. Write the Lagrangian of the particle in spherical coordinates.
2. Calculate the conjugate momenta p_r, p_θ, and p_ϕ.
3. Show that the momentum p_ϕ is equal to the z component of the angular momentum L_z whose expression in Cartesian coordinates is $L_z = xp_y - yp_x$.
4. To what invariance law does the conservation of L_z correspond?
5. If the particle is charged and placed in a magnetic field $\mathbf{B}$ parallel to the z axis, is the component L_z conserved?

3.4. Brachistochrone

A popular problem for mathematicians of the 17th century was the *brachistochrone* curve. Consider two points A and B in a vertical plane, joined by a curve C. In A, a massive particle is dropped with zero initial velocity, and it slides without friction along the curve under the effect of gravity. We want to determine the curve C such that the time for the particle to go from A to B is minimum. We note z the altitude and x the abscissa of a point on the curve. The endpoints A and B correspond respectively to $(x = a, z = \alpha)$ and $(x = b, z = \beta)$.

3.6   Problem. Strategy of a Regatta

A sailboat has velocity $v(\theta)$, which is a function of the angle θ between the direction of the wind and the direction of the boat and also of the norm w of the velocity of the

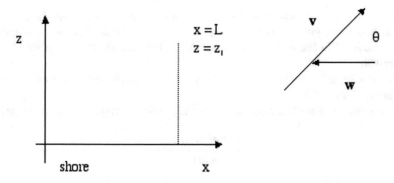

Fig. 3.2 Diagram of the direction of the sailboat v compared with that of the wind w

wind. We assume that the velocity of the boat v is proportional to the velocity of the wind w and that it depends on the angle θ chosen by the skipper. For convenience, in what follows, we shall write this velocity in the form

$$v(\theta) = \frac{w}{\cos(\theta)\, h(\tan\theta)}, \quad \text{with } h(u) = \frac{1}{2}\left(u + \frac{1}{u}\right). \quad (3.86)$$

We are interested in the strategy where the sailboat tacks to the wind (i.e., $\theta \leq \pi/2$), as shown in Fig. 3.2. We assume that the x component v_x of the velocity of the boat is opposite to that of the wind and that the position of the sailboat along the x axis always increases with time. We assume the coast is linear (land = half-plane $z < 0$, sea = half-plane $z > 0$).

We assume the wind is parallel to the coast, of direction opposite to the x axis, and that the norm of its velocity $w(z)$ depends only on the distance z to the coast.

Here, we assume that the velocity of the wind has the form

$$w(z) = w_0 - w_1 \frac{z_0}{z + z_0}, \quad (3.87)$$

where w_0 is the velocity far from the coast, which is larger than the velocity $(w_0 - w_1) \geq 0$ on the coast $z = 0$.

1. We denote
$$\dot{x} = \frac{dx}{dt}, \quad \dot{z} = \frac{dz}{dt}, \quad z' = \frac{dz}{dx}.$$

Show that $z' = \tan\theta$.
2. We first assume the wind is uniform ($w = $ constant, $w_1 = 0$). Write the expression of the velocity of the boat along the axis of the wind $v_x = \dot{x}$ in terms of w and $h(\tan\theta)$. For what values of θ and z' is this velocity maximum? What is its value?
3. We now assume $w_1 \neq 0$. The boat sails from the origin ($x = 0$, $z = 0$) to a given point ($x = L$, $z = z_1$). We assume that $z' \geq 0$ for all t (i.e. the boat never changes

tack). We want to determine the fastest trajectory $z(x)$. Write the expression of the time dt to go, on this trajectory, from x to $x + dx$ in terms of the functions w and h. Give the value of the total time T between the starting point and the arrival.

4. Deduce from (3) the equation that determines the optimal trajectory (which minimizes T).

5. Show that the translation invariance of the problem along the x direction yields

$$\frac{h'(z')z' - h(z')}{w(z)} = A,$$

where A is a constant.

6. Use the previous result to calculate the trajectory in the form of a function $x(z)$ (and not a function $z(x)$). Fix the value of the constant A.

7. Calculate the value of $z' = dz/dx$ as a function of z. We assume that $z_1 \ll L$ and $z_1 \ll z_0$. Do you think the result corresponds to the best strategy? If not, what modifications must the skipper make?

Chapter 4
Hamilton's Canonical Formalism

It is in the silence of laws that great actions are born.
Donatien, Alphonse, Marquis de Sade

The work of Lagrange was followed by the monumental, five volume, *Traité de Mécanique Céleste* (Treatise of Celestial Mechanics) of Laplace, published between 1799 and 1825. This work had a crucial importance both in Astronomy and in the evolution of philosophical ideas.

This leads us to the next step, in the 1830's, and to the so-called *canonical formalism* of Analytical mechanics due to Hamilton.[1]

Hamilton's canonical formalism was elaborated in 1834.[2] It is more convenient for a series of problems such as the dynamics of point-like particles. But it is impressive, above all, by the number of its developments, both in physics and in mathematics. In the present book, we are mainly concerned with applications to mechanics, but we shall describe several other spin-offs of Hamilton's work. This work is actually of an impressive richness by its completely new mathematical and physical developments, concerning to the *symmetry* properties, which it reveals and which allow us to discover unexpected properties and to tackle entirely new problems.

Hamilton had understood that, unlike intuitive variables, coordinates and their time derivatives, coordinates and *conjugate momenta* have *symmetric* dynamical

---

[1] As in the previous chapter, one can refer to Herbert Goldstein [3] or to Landau and Lifshitz [1] and [2], for any development which would be missing here.

[2] This formalism was developed by Hamilton in: *On a General Method in Dynamics*, Philosophical Transactions of the Royal Society, part II for 1834, pp. 247–308, in a brilliant form.

© The Author(s), under exclusive license to Springer Nature Switzerland AG 2023   63
J.-L. Basdevant, *Variational Principles in Physics*,
https://doi.org/10.1007/978-3-031-21692-3_4

roles in Lagrangian mechanics, and this gives a lot of depth both to the problems considered and to their solution.

This text is primarily oriented towards applications to mechanics, but we will end with another consequence of Hamilton's work, *dynamical systems*, whose founder was Henri Poincar (1854–1912).

It is in the next chapter that we will describe more fully Hamiltons complete work, which goes from optics to mathematics: in particular, he invented quaternions.

In Sect. (4.1), we explain this canonical formalism which consists in describing the state of a system by the conjugate variables positions $\{x\}$ and Lagrange conjugate momenta $\{p\}$, and not by positions and velocities. In other words, a system is described by a point in *phase space*, its is governed by a *Hamiltonian* which is obtained from the Lagrangian by a Legendre transformation.

In Sect. (4.2), we present the *Poisson brackets*, an important new mathematical structure. Jacobi considered them as Poisson's greatest discovery. In fact Poissons brackets bear the germ of the group theory of Sophus Lie (1842–1899). This study allows us to have a new vision of conservation laws.

At this point, we will arrive at a discovery of Dirac in 1925. There is a perfect *structural symmetry between analytical and quantum mechanics*, if we match the Poisson brackets of the classical physical quantities, and the *commutators* of quantum observables (divided by $i\hbar$).

In Sect. (4.3), we will define the *canonical transformations*, which have many applications, and which exemplify the equivalence between the state variables $\{x\}$ and $\{p\}$.

This will lead us, in Sect. (4.4), to the *phase space* which is, from the mathematical point of view, the actual space appropriate to the description of the evolution of a system of points, contrary to the empirical space of position and velocity variables. We will establish in particular the *Liouville theorem* (1809–1882), geometric property of the *Hamiltonian flow* of the trajectories of a system in its phase space. We will see the remarkable geometric structure of the evolution of a dynamic system, a structure that cannot appear with the classical variables x, $\dot{x}$.

In Sect. (4.5), we will come back to the case of a charged particle in a magnetic field treated in Sect. (3.3.4) of the previous chapter, where precisely the conjugate momentum and the classical momentum differ radically. In quantum mechanics, it has been demonstrated *experimentally* the fact that the Hamiltonian is expressed in terms of the potentials, not the fields.

Finally, after writing Hamilton's canonical equations, which are first order coupled differential equations of the evolution of the state state variables, we shall present, in Sect. (4.6), some aspects of *dynamical systems*. In fact, this type of physical problem was an amazing source of discoveries, both in mathematics and in physics. Henri Poincaré founded this field of research in 1885 when he studied the three-body problem. This leads to fascinating developments, such as the behavior for $t = \infty$, attractors and strange attractors, bifurcations, chaos etc. The most famous strange attractor is the Lorenz attractor, after its inventor, Edward N. Lorenz, who discovered it in 1963 in a mathematical model of the evolution of the atmosphere.

Lorenz gave a new and spectacular source of interest to chaos with his "butterfly" effect in meteorology.

4.1 Hamilton's Canonical Formalism

Actually, the formulation (3.2) of the least action principle is not due to Lagrange (who used a more complicated form). It is due to Hamilton in 1834. Hamilton, one of the greatest figures of science, was fascinated by Lagrange and by his analytical mechanics which he called a "scientific poem written by the Shakespeare of mathematics".

Hamilton's canonical formalism was formulated in 1834. It is more convenient for some categories of problems, but it goes far beyond that. It contains a particularly fruitful mathematical structure which leads to Lie groups, to dynamical systems, and many other developments.

Hamilton's starting point is to describe the state of a system by the variables x_i, positions, and p_i, the conjugate momenta, instead of x_i and $\dot{x}_i$.

4.1.1 Canonical Equations

Suppose that we invert Eq. (3.12) and that we can calculate the $\{\dot{x}_i\}$ in terms of the $\{x_i\}$ and $\{p_i\}$, which are our new state variables.[3]

The problem is to obtain the equations of motion of the $\{x_i\}$ and $\{p_i\}$ in terms of these same variables, by eliminating the $\{\dot{x}_i\}$.

The solution consists in performing a Legendre transformation. Let us introduce the Hamilton function or *Hamiltonian*

$$H(\{x_i\}, \{p_i\}, t) = \sum_i p_i \dot{x}_i - \mathcal{L}. \tag{4.1}$$

Consider, for simplicity, a one-dimensional problem. The total differential of H is

$$dH = p\, d\dot{x} + \dot{x}\, dp - \frac{\partial \mathcal{L}}{\partial x} dx - \frac{\partial \mathcal{L}}{\partial \dot{x}} d\dot{x} - \frac{\partial \mathcal{L}}{\partial t} dt.$$

Taking into account (3.12) and (3.13), the first and fourth terms cancel, and the third one is $-\dot{p}\, dx$. Therefore, we have

$$dH = \dot{x}\, dp - \dot{p}\, dx - \frac{\partial \mathcal{L}}{\partial t} dt, \tag{4.2}$$

---

[3] Conjugate momenta always exist since the Lagrangian contains a quadratic term in $\dot{x}_i$.

which provides us with the equations of motion

$$\dot{x} = \frac{\partial H}{\partial p}, \qquad \dot{p} = -\frac{\partial H}{\partial x}. \tag{4.3}$$

For a system with several degrees of freedom, we have

$$\dot{x}_i = \frac{\partial H}{\partial p_i}, \qquad \dot{p}_i = -\frac{\partial H}{\partial x_i} \tag{4.4}$$

which are called the *canonical equations* of Hamilton.

> Legendre transformations are often used for performing changes of variables. One chooses the most convenient set of variables according to the nature of the physical problem under consideration.
>
> A simple example is that of thermodynamic potentials. Starting from the energy $U = W + Q$ which is convenient if one works with the volume and the entropy, $dU = -PdV + TdS$, one goes to the enthalpy $H = U + PV$ if one works with the pressure and the entropy $dH = VdP + TdS$, to the free energy $F = U - TS$ if one works with the volume and the temperature $dF = -PdV - SdT$, and the free enthalpy, the Gibbs function $G = F + PV$, if one works with the temperature and the pressure $dG = -SdT + VdP$.

Hamilton's equations (4.4) form a set of *first order* coupled differential equations in the time variable, which is a major advantage. They are symmetric in x and p (up to the minus sign, which we shall come back to). They possess the major technical advantage to present directly the time evolution of the state variables in terms of these same variables.

The value of the Hamilton function is, of course the energy (3.18). If the Lagrangian does not depend explicitly on time, $\partial L/\partial t = 0$, then $\partial H/\partial t = 0$ and the energy is conserved

$$\frac{\partial H}{\partial t} = 0 \longrightarrow \frac{d}{dt}H = 0. \tag{4.5}$$

4.2 Poisson Brackets, Phase Space

Consider two physical quantities f and g, which are functions of the state variables (x_i, p_i), $i = 1, \ldots, N$ and possibly of time. One calls *Poisson bracket* of f and g the quantity

$$\{f, g\} = \sum_{i=1}^{N} \left(\frac{\partial f}{\partial x_i} \frac{\partial g}{\partial p_i} - \frac{\partial f}{\partial p_i} \frac{\partial g}{\partial x_i} \right). \tag{4.6}$$

Poisson brackets have the following properties, which are straightforward to establish

$$\{f, g\} = -\{g, f\}, \qquad \{f_1 + f_2, g\} = \{f_1, g\} + \{f_2, g\} \tag{4.7}$$

$$\{f_1 f_2, g\} = f_1\{f_2, g\} + \{f_1, g\}f_2. \tag{4.8}$$

For the state variables (x_i, p_i) we have the important relations

$$\{x_i, x_j\} = 0 \quad \{p_i, p_j\} = 0 \quad \{x_i, p_j\} = \delta_{ij}, \tag{4.9}$$

and

$$\{x_i, f\} = \frac{\partial f}{\partial p_i} \quad \{p_i, f\} = -\frac{\partial f}{\partial x_i}. \tag{4.10}$$

One obtains with no difficulty the Jacobi identity

$$\{f, \{g, h\}\} + \{g, \{h, f\}\} + \{h, \{f, g\}\} = 0. \tag{4.11}$$

4.2.1 Time Evolution, Constants of the Motion

We now calculate the time evolution of a physical quantity $f(x_1, p_1, \ldots, x_N,$ $p_N; t) \equiv f([x_i, p_i]; t)$, where we note $[x_i, p_i]$ the set of variables $(x_1, p_1, \ldots, x_N, p_N)$ in order to avoid any confusion with Poisson brackets. We have

$$\dot{f} = \frac{df}{dt} = \sum_i \left(\frac{\partial f}{\partial x_i}\dot{x}_i + \frac{\partial f}{\partial p_i}\dot{p}_i\right) + \frac{\partial f}{\partial t}. \tag{4.12}$$

Using Hamilton's equations (4.4), we obtain

$$\dot{f} = \{f, H\} + \frac{\partial f}{\partial t}. \tag{4.13}$$

In particular, the canonical equations (4.4) are now written in a *symmetric* way

$$\dot{x}_i = \{x_i, H\} \quad \dot{p}_i = \{p_i, H\}. \tag{4.14}$$

In the canonical formalism, the Hamiltonian governs the time evolution of the system. If a physical quantity f does not depend explicitly on time, i.e. $\partial f/\partial t = 0$ (which amounts to saying that the system is isolated) then its time evolution is obtained by taking the Poisson bracket of f and the Hamiltonian

$$\dot{f} = \{f, H\}. \tag{4.15}$$

Therefore, if the Poisson bracket of a quantity f and the Hamiltonian vanishes, f is a *constant of the motion*.

The Poisson brackets, which were invented in 1809, are far from being simple technical tools. Jacobi (1804–1851) considered them as Poisson's greatest discovery.

In fact, the Poisson brackets contain the germ of the group theory of Sophus Lie (1842–1899).

Poisson Theorem

Theorem 1 *If f and g are two constants of the motion, then their Poisson bracket is also a constant of the motion.*

This theorem, due to Poisson, can be derived from the Jacobi identity (4.11)

$$\{H, \{f, g\}\} + \{f, \{g, H\}\} + \{g, \{H, f\}\} = 0. \tag{4.16}$$

We assume that f and g are constants of the motion, i.e. $\{g, H\} = 0$ and $\{H, f\} = 0$. Therefore,

$$\{H, \{f, g\}\} = 0$$

and $\{f, g\}$ is a constant of the motion. In certain cases, this allows one to find new constants of the motion.

> Jacobi wrote, with sharpness: *[this is] Mr. Poissons deepest discovery, which, I believe, was not well understood neither by Lagrange, nor by the many others who quoted it, nor by its author himself. The theorem I am talking about seems to me the most important in mechanics [...] this truly prodigious theorem remained at the same time discovered and hidden.*

Rotations and Poisson Brackets of Angular Momentum

The three components of the angular momentum $\mathbf{L} = \mathbf{r} \times \mathbf{p}$ are connected by the Poisson brackets

$$\{L_x, L_y\} = L_z \quad \{L_y, L_z\} = L_x \quad \{L_z, L_x\} = L_y. \tag{4.17}$$

This reflects the 3-dimensional group structure of rotations, $SO(3)$. If the Poisson bracket of a Hamiltonian with the angular momentum cancels, that Hamiltonian is rotation invariant.

4.2.2 Relation Between Analytical and Quantum Mechanics

The above formulas reveal an amazing fact. There is a strong analogy, if not more, between the structures of analytical mechanics and of quantum mechanics. In quantum mechanics, one proves quite easily what is called the Ehrenfest theorem:[4] the

---

[4] See for instance [11] Chap. 7, Sect. 3.

time derivative of the expectation value $\langle a \rangle$ of a physical quantity A is related to the commutator of the observable $\hat{A}$ and the Hamiltonian $\hat{H}$ by the relation

$$\frac{d}{dt}\langle a \rangle = \frac{1}{i\hbar}\langle[\hat{A}, \hat{H}]\rangle + \langle\frac{\partial \hat{A}}{\partial t}\rangle. \tag{4.18}$$

If, by definition, we introduce an operator $\hat{\dot{A}}$ such that whatever the state vector $|\psi\rangle$ we have

$$\langle\psi|\hat{\dot{A}}|\psi\rangle := \frac{d}{dt}\langle a \rangle,$$

then we have the equality of observables

$$\hat{\dot{A}} = \frac{1}{i\hbar}[\hat{A}, \hat{H}] + \frac{\partial \hat{A}}{\partial t}. \tag{4.19}$$

which has the same structure as (4.13) if one replaces the Poisson brackets by the commutators, divided by $i\hbar$, of the quantum observables.

The same remark applies to the canonical commutation relations of the conjugate variables of position $\hat{x}$ and momentum $\hat{p}$

$$[\hat{x}_j, \hat{p}_k] = i\hbar\delta_{jk} \tag{4.20}$$

which can be compared with (4.9).

This similarity, if not identity, between the structures of the two mechanics was one of the first major discoveries of Dirac during the summer of 1925 (he was 23). Dirac, after finding that the non-commutativity of quantum observables was actually the founding stone of Heisenberg's matrix mechanics, had decided to construct a new formulation of mechanics which would incorporate this non-commutativity in a well defined way. One day, he remembered the structure of Poisson brackets and he saw that they played, formally, a similar role as the quantum commutators, divided by $i\hbar$. On september 1925 he was able to construct what is called "Quantum mechanics" in its present form. Of course the mathematical nature and the physical interpretation of the quantities are different in the two cases, but the equations which relate them are the same if one postulates the correspondence between Poisson brackets in analytical mechanics and the quantum commutators, divided by $i\hbar$, in quantum mechanics.

More generally, in complex problems (large numbers of degrees of freedom, constraints between variables, etc.) the systematic method to obtain the commutation relations of quantum observables consists in referring to the classical Poisson brackets and in replacing them by the form of quantum commutators.

4.3   Canonical Transformations in Phase Space

In the Lagrangian formalism, the Lagrange-Euler equations keep the same form in any change of coordinates $x_i \longrightarrow X_i(x_1, \ldots, x_n)$ (for instance, changing from Cartesian coordinates (x, y, z) to polar coordinates (r, θ, ϕ)). These changes of coordinates in configuration space are called point transformations.

In the Hamiltonian formalism there exists a much larger class of transformations under which the equations of motion are invariant. One can, indeed, mix the state variables, i.e. the positions $\{x_i\}$ and the conjugate momenta $\{p_i\}$, and perform a transformation in *phase space*, whose interest we see below. One calls *canonical transformation* a coordinate transformation

$$X_i(x_1, \ldots, x_N, p_1, \ldots, p_N; t) , \quad P_i(x_1, \ldots, x_N, p_1, \ldots, p_N; t) \qquad (4.21)$$

such that Hamilton's equations keep the same form in the new variables. Let $H'(X_1, \ldots, X_N, P_1, \ldots, P_N; t)$ be the Hamilton function expressed in terms of the new variables $[X_i, P_i]$, then, in a canonical transformation, by definition one has

$$\dot{X}_i = \frac{\partial H'}{\partial P_i} , \quad \dot{P}_i = -\frac{\partial H'}{\partial X_i}. \qquad (4.22)$$

The following theorem is of great practical importance.

Theorem 2 A transformation $[x_i, p_i] \to [X_i, P_i]$ which preserves the Poisson brackets is a canonical transformation. It preserves the equations of motion.

This amounts to requiring that the Poisson brackets expressed in terms of the new variables be the same as those expressed in the initial variables, i.e.

$$\{X_i, X_j\} = 0 \quad \{P_i, P_j\} = 0 \quad \{X_i, P_j\} = \delta_{ij}. \qquad (4.23)$$

We can give a direct proof. In order to simplify things, consider a time-independent transformation and a single couple of variables $(x, p) \to (X(x, p), P(x, p))$ such that $\{X, P\} = 1$. We note $H(x, p)$ and $H'(X, P)$ the expression of Hamilton's function in these two systems of variables.

In the variables, (x, p), the time evolution of X and P is $\dot{X} = \{X, H\}$ and $\dot{P} = \{P, H\}$, which can be written, for instance, as

$$\dot{X} = \frac{\partial X}{\partial x} \frac{\partial H}{\partial p} - \frac{\partial X}{\partial p} \frac{\partial H}{\partial x}. \qquad (4.24)$$

The Hamilton function in the new variables is expressed as

$$H'(X, P) = H(x(X, P), p(X, P)) \quad , \qquad (4.25)$$

and its inverse as

$$H(x, p) = H'(X(x, p), P(x, p)) \ . \tag{4.26}$$

Differentiating H with respect to x and p in the previous expression, we obtain

$$\frac{\partial H}{\partial x} = \frac{\partial H'}{\partial X}\frac{\partial X}{\partial x} + \frac{\partial H'}{\partial P}\frac{\partial P}{\partial x} \quad ; \quad \frac{\partial H}{\partial p} = \frac{\partial H'}{\partial X}\frac{\partial X}{\partial p} + \frac{\partial H'}{\partial P}\frac{\partial P}{\partial p}. \tag{4.27}$$

Using this in (4.24), we obtain

$$\dot{X} = \left(\frac{\partial X}{\partial x}\frac{\partial P}{\partial p} - \frac{\partial X}{\partial p}\frac{\partial P}{\partial x}\right)\frac{\partial H'}{\partial P} \equiv \{X, P\}\frac{\partial H'}{\partial P},$$

$$\dot{P} = \left(\frac{\partial P}{\partial x}\frac{\partial X}{\partial p} - \frac{\partial P}{\partial p}\frac{\partial X}{\partial x}\right)\frac{\partial H'}{\partial X} \equiv -\{X, P\}\frac{\partial H'}{\partial X}.$$

Since, by assumption, $\{X, P\} = 1$, we indeed recover the canonical equations

$$\dot{X} = \frac{\partial H'}{\partial P}, \quad \dot{P} = -\frac{\partial H'}{\partial X} \quad QED. \tag{4.28}$$

(a) Comments

1. The extension to an arbitrary number N of variables

$$(x_1, \ldots, x_N, p_1, \ldots, p_N) \rightarrow (X_1, \ldots, X_N, P_1, \ldots, P_N)$$

with by assumption $\{X_i, X_j\} = 0$, $\{P_i, P_j\} = 0$, $\{X_i, P_j\} = \delta_{ij}$ causes no diffi-culty (see [1]).
2. We see that since canonical transformations mix coordinates and momenta, there is no fundamental difference between theses two types of state variables. In the Hamiltonian formalism, the notions of space coordinates and momenta (more or less associated with linear momenta) loose their intuitive meaning.
   For this reason, one usually calls these variables *canonical conjugate variables* which are noted as (q_i, p_i) with the relations $\{q_i, p_j\} = \delta_{ij}$, $\{q_i, q_j\} = \{p_i, p_j\} = 0$. The very simple canonical transformation ($X = p$, $P = -x$) shows that, in that sense, these two variables can be "exchanged". The canonical conjugate variables characterize the state of the system by a point in *phase space* (see below).
3. The Hamiltonian motion, that is the evolution of the system, therefore appears *at each time t* as performing a *canonical transformation* of the state variables

$$(x_i(t), p_i(t)) \rightarrow (x_i(t'), p_i(t')) \quad \text{with} \quad \{x_i(t), p_i(t)\} : \delta_{ij} \quad \forall t'.$$

4. In general, one calls *canonical conjugate variables* q and p two physical quantities such that $\{q, p\} = 1$. One example, in spherical coordinates, is the azimuthal angle φ and the z component of the angular momentum L_z (see exercise (3.3) of Chap. 2).

(b) Example: the one-dimensional harmonic oscillator, angle-action variables.

Consider a one-dimensional harmonic oscillator of Hamiltonian $H = p^2/(2m) + m\omega^2 x^2/2$ where x and p are canonical conjugate variables. The transformation $x = X/\sqrt{m\omega}$ and $p = P\sqrt{m\omega}$ is a canonical transformation: $\{X, P\} = 1$, and, with these variables, the Hamiltonian is written as $H = \omega(P^2 + X^2)/2$. The rotation in phase space

$$\xi = X\cos\theta + P\sin\theta , \quad \Pi = P\cos\theta - X\sin\theta, \tag{4.29}$$

where θ is an arbitrary fixed angle, is a canonical transformation. The expression of Hamilton's function is of the same form : $H = \omega(\Pi^2 + \xi^2)/2$ and $\{\xi, \Pi\} = \{X, P\} = 1$.

This is a simple but important example of a *dynamical symmetry* of the system. In the present case, it is one example of the numerous symmetry properties of the harmonic oscillator. This argument can be extended to N degrees of freedom.

Dirac's method of creation and annihilation operators in quantum mechanics[5] relies directly on this symmetry.

(c) Cyclic variables

This symmetry can further be exploited. In phase space, which is two-dimensional, (X, P), in the above example, we can use polar coordinates by introducing the variables (A, φ) defined by

$$X = \sqrt{2A}\cos\varphi, \quad P = \sqrt{2A}\sin\varphi, \tag{4.30}$$

which amounts to

$$A = \frac{X^2 + P^2}{2}, \quad \varphi = \arctan(\frac{P}{X}). \tag{4.31}$$

The variables (A, φ) are canonical conjugate variables, as one can check with no difficulty. In these variables, the Hamiltonian reduces to the simple expression $H = \omega A$. Hence the equations of motion

$$H = \omega A, \quad \{A, \varphi\} = 1 \Rightarrow \dot{A} = 0 , \quad \dot{\varphi} = \omega, \tag{4.32}$$

whose solution is obvious

$$A = E/\omega = \text{constant}, \quad \varphi = \omega t + \varphi_0. \tag{4.33}$$

Here, E is the energy of the oscillator, a constant of the motion. The interesting point about this operation is that we have reduced the problem to a single time-dependent variable, the angle φ. Since the energy, which is proportional to the *action A*, is

---

[5] See for instance [11] Chap. 7, Sect. 5.

conserved, only the angular variable φ evolves. The variable φ is a *cyclic variable*. It does not appear explicitly in the Hamiltonian, and this results in the properties (4.32) and (4.33).

The geometric interpretation in the (X, P) space, which is here equivalent to phase space, is simple. The motion occurs on a circle of radius $A = E/\omega$ which depends on the energy E, constant of the motion. On this circle, the motion of the point (X, P) is uniform, of angular velocity $\omega : \varphi = \omega t + \varphi_0$.

We already mentioned cyclic variables in Sect. (3.2.2). This is a simple example of the role played such variables in particular in the investigation of integrable systems.

Coupled Oscillators

The case, more suggestive geometrically, of *two identical coupled oscillators*

$$H = \frac{1}{2}(p_1^2 + k^2 x_1^2) + \frac{1}{2}(p_2^2 + k^2 x_2^2) + \frac{1}{2}K^2(x_1 - x_2)^2$$

separates by the canonical transformation

$$P = (p_1 + p_2)/\sqrt{2}, \; X = (x_1 + x_2)/\sqrt{2}, \quad Q = (p_1 - p_2)/\sqrt{2}, \; Y = (x_1 - x_2)/\sqrt{2}$$

into

$$H = H_1 + H_2 \quad \text{with} \quad H_1 = \frac{1}{2}(P^2 + \omega_1^2 X^2) \quad \text{and} \quad H_2 = \frac{1}{2}(Q^2 + \omega_2^2 Y^2)$$

where $\omega_1 = k$ and $\omega_2 = \sqrt{k2 + K2}$.

We then obtain, in phase space, a toric motion, periodic or not depending on whether ω_1/ω_2 is rational or not, represented in Fig. (4.1b)

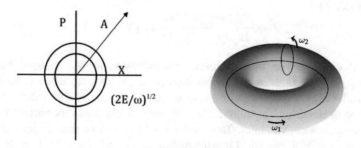

Fig. 4.1 a–Left: Path of the state of a one-dimensional harmonic oscillator in phase space. **b**–Right: Dynamics on a torus, in phase space, of two coupled oscillators

Note that in this case the total energy of the system can be written as $E = E_1 + E_2$ where the two energies E_1 and E_2 are constants of the motion. If we refer to (4.33), we see that the relative scale depends on E_1 and E_2, that are determined by the initial conditions.

4.4 Evolution in Phase Space: Liouville's Theorem

The evolution in *phase space* $[x_i, p_i]$ is a geometrical representation of particular interest in mechanics. A point in phase space corresponds to a state of the system. When the system evolves, this point moves in phase space. A volume element of phase space is defined by

$$d\Omega = dx_1 \ldots dx_N \, dp_1 \ldots dp_N. \tag{4.34}$$

Consider an arbitrary volume Ω of phase space, $\Omega = \int d\Omega$, we claim that this volume is invariant under canonical transformations

$$\int dx_1 \ldots dx_N \, dp_1 \ldots dp_N = \int dX_1 \ldots dX_N \, dP_1 \ldots dP_N. \tag{4.35}$$

Indeed, in this change of variables, we have

$$\int dX_1 \ldots dX_N \, dP_1 \ldots dP_N = \int |J| dx_1 \ldots dx_N \, dp_1 \ldots dp_N$$

where J is the Jacobian of the transformation. Now, the Jacobian of a canonical transformation is equal to one.

This is obvious on the simple example (4.29) above.

If we consider the simple case of a single couple of conjugate variables $(x, p) \rightarrow (X, P)$ as in (4.24), the proof is simple. Indeed, the Jacobian is simply the Poisson bracket $\{X, P\}$

$$J = \frac{\partial X}{\partial x} \frac{\partial P}{\partial p} - \frac{\partial X}{\partial p} \frac{\partial P}{\partial x} = \{X, P\} = 1 \tag{4.36}$$

which is equal to one by definition. The extension to N conjugate variables $x_1 \ldots x_N \, p_1 \ldots p_N$ is more lengthy but it proceeds from the same observation.

Consider now a volume Ω of phase space. Each point of this volume evolves according to Hamilton's equations. Consequently, at any time t the variables $(x_i(t), p_i(t))$ are canonical variables. Therefore the motion can be seen as performing at each time interval a canonical transformation of the state variables in phase space $(x_i(t), p_i(t)) \rightarrow (x_i(t'), p_i(t'))$. We therefore obtain the *Liouville theorem*, of great importance in statistical physics:

Theorem 3 *A volume of phase space remains unchanged in the Hamiltonian evolution of the system.*

This remarkable geometric property derives from the structure of Hamilton's equations. It is independent of the specific form of the Hamiltonian itself.

Another interesting geometric property in phase space is the following. Hamilton's function $H(x, p)$ is defined in phase space. In this space, consider a vector field whose components are $(\dot{x}, \dot{p})$, i.e.

$$\dot{x} = \frac{\partial H}{\partial p}, \quad \dot{p} = -\frac{\partial H}{\partial x}.$$

One calls the *flow* of this vector field the set of curves whose tangents at each point are collinear with the vector at this point. We notice that the flow of $(\dot{x}, \dot{p})$, also called the Hamiltonian flow, is orthogonal in each point to the *gradient* of the Hamiltonian at this point

$$\nabla H = \left(\frac{\partial H}{\partial x}, \frac{\partial H}{\partial p} \right).$$

In the example (4.29) above, the result is very simple. The trajectories in the (X, P) plane are circles centered at the origin, and the gradient of $H = (P^2 + X^2)/2$ lies along straight lines going through the origin. This can be stated in the reverse way : the gradient of $H = (P^2 + X^2)/2$ lies along straight lines going through the origin, therefore the trajectories are circles centered at the origin. This result can be generalized to any number of variables. One can express the conservation laws of energy, momentum and angular momentum with geometrical considerations of that kind (using the corresponding invariance properties).

4.5 Charged Particle in an Electromagnetic Field

Consider the form (3.52) of the Lagrangian of a charged particle placed in an electromagnetic field

$$\mathcal{L} = \frac{1}{2}m\dot{\mathbf{r}}^2 + q\,\dot{\mathbf{r}} \cdot \mathbf{A}(\mathbf{r}, t) - q\,\Phi(\mathbf{r}, t). \tag{4.37}$$

The conjugate momentum is

$$\mathbf{p} = m\dot{\mathbf{r}} + q\mathbf{A}(\mathbf{r}, t). \tag{4.38}$$

4.5.1 Hamiltonian

Equation (4.38) is easily inverted $\dot{\mathbf{r}} = (\mathbf{p} - q\mathbf{A}(\mathbf{r}, t))/m$, hence the Hamiltonian

$$H = \frac{1}{2m}(\mathbf{p} - q\mathbf{A}(\mathbf{r}, t))^2 + q\Phi(\mathbf{r}, t) \tag{4.39}$$

Which is expressed in terms of the *potentials* $\mathbf{A}$ and Φ, and not the fields $\mathbf{E}$ and $\mathbf{B}$.

As an exercise, one can write the Hamiltonian in the relativistic case, (3.72), the result is

$$H = \sqrt{m^2 c^4 + c^2(\mathbf{p} - q\mathbf{A})^2} + q\Phi \tag{4.40}$$

where one discovers the "prescription" to introduce the electromagnetic field. One must substitute $\mathbf{p} - q\mathbf{A}$ to the momentum $\mathbf{p}$ and $E + q\phi$ to the energy E in the expression of the energy-momentum relation for a free particle.

It is that prescription which Schrödinger applied to the free wave equation for de Broglie waves in order to calculate the energy levels of the hydrogen atom. After some unexpected mismatches, he finally ended up with his celebrated equation (to leading order in v/c the hydrogen atom spectrum is nonrelativistic).

4.5.2 Gauge Invariance

As in Chap. 3, Sect. (3.3.3), we see that the hamiltonian is expressed in terms of the potentials and not in terms of the fields themselves. Contrary to what happens in (3.3.3), it is not obvious *a priori* that the result will be gauge invariant. One can check this directly on the equations of motion, of course.

In the expression (4.39) of the hamiltonian, the quantity $(\mathbf{p} - q\mathbf{A})/m$ is the *velocity* of the particle. It is a measurable quantity (some m.s^{-1}), independent of the gauge. Similarly, the energy is $E = mv^2/2 + q\Phi$ as it should. However, the conjugate momentum $\mathbf{p}$ depends on the gauge.

One can see[6] how quantum mechanics deals with the subtle problems of gauge invariance, which leads directly to the basic principles which dictate the form of fundamental interactions. One can also devise experimental setups which show directly that the Hamiltonian is expressed as a function of the potentials and not the fields. An example can be found in the Aharonov and Bohm experiment.[7]

---

[6] See for instance [11], Chap. 15, Sect. 5.

[7] A. Tonomura et al., Phys. Rev. Lett. **56**, 792, (1986).

4.6 Dynamical Systems

More generally, if we note $\mathbf{X}(t) = (\mathbf{r}_i(t), \mathbf{p}_i(t))$ the position of the system at time t in *phase space*, Hamilton's equations are of the form $\dot{\mathbf{X}} = F(\mathbf{X})$, i.e. a first order differential equation for the evolution of the $2N$-component vector $X(t)$. This is called a *dynamical system*.

This seems relatively simple when the number of components is small. In fact, this was the case in the early development of mechanics with the discoveries of Galileo, Newton and all the mathematicians and physicists who built the evolution in time of fairly simple systems, even in astronomy (Kepler's ellipses, Halley's comet etc.). Statistical physics, on the other hand, has grown thanks to the methods developed as a result of Boltzmanns work. But many domains seem out of reach of computers as soon as the number of variables is too large.

This type of problem has been an important source of discoveries both in mathematics and in physics; one can refer to the book of I. Percival and D. Richards [13] and to the works of D. Ruelle and I. Ekeland.[8]

In the case of large dimensions, certainly unattainable by the techniques of differential calculus, this type of problem has been an incredible source of discoveries both in mathematics and physics and in many fields, by studying what are called *Dynamical Systems*. The evolution over time of complex systems has been and is at present one of the very great problems both in basic research, in mathematics, and in applied research. What is important is not to obtain a numerical precision on a given quantity, but to identify qualitatively and quantitatively astonishing global behaviors that a sequence of numbers can hardly express.

4.6.1 The Contribution of Henri Poincaré

The founder of this field of investigations is Henri Poincaré[9] in an impressive series of publications[10] especially in his study of the three body problem (the really difficult problem of mechanics) in 1885 and after. A large number of mathematicians followed this type of problem, and obtained often unexpected results of considerable practical importance.

One studies the whole set of possible motions which is called the *flow* of these vectors. This leads to fascinating problems such as limiting problems at $t = \infty$,

---

[8] David Ruelle, "Turbulence, Strange Attractors, and Chaos", World Scientific. 16: 195, (1995); Ivar Ekeland, *The broken dice, and other mathematical tales of chance*, University of Chicago Press. pp. iv+183. ISBN 978-0-226-19991-7 (1993).

[9] See for example Charpentier Eric, Ghys Etienne, Lesne Annick editors (2010). *The scientific legacy of Poincaré*, collection "History of Mathematics", American Mathematical Society, Providence, RI, ISBN: 978-0-8218-4718-3.

[10] *The new methods of celestial mechanics. Integral invariants; second-type periodic solutions; doubly asymptotic solutions* by H. Poincaré, 1892–1899, https://gallica.bnf.fr/ark:/12148/bpt6k96109954.

attractors and strange attractors; bifurcations, which are sudden changes in the nature of these flows for certain values of the parameters entering the function $F(\mathbf{X})$, chaos and the "butterfly effect" in meteorology, etc.

4.6.2   Poincaré and Chaos in the Solar System

Poincaré proved that in a gravitating system involving more than two bodies, say planets around the sun, to initial conditions as close as one wishes, there corresponds a time when two of the planets can be as far away from each other (and from their starting point) as one wishes. In the 19th century, Laplace and others had extensively developed perturbation theory which provided extremely accurate predictions for celestial mechanics. However, in the course of his work, Poincaré showed that the perturbation expansion never converges; it is only an asymptotic expansion of which only the first terms are useful and can be used over a finite time interval. This effect is called *chaos*. It manifests itself by the fact that two solutions corresponding to extremely close initial conditions, differ by a quantity increasing with time as $exp(t/\tau)$, where the characteristic time τ, called the Liapounov horizon, depends on the problem considered. It occurs in many other physical problems. Depending on the system considered, Liapounovs horizon is very variable, it is about 200 million to one billion years for the solar system.[11]

A very simple example of a chaotic system is playing dice. In principle, in classical mechanics, if we were to determine extremely accurately the conditions of the problem (i.e. the initial conditions, the way to throw the dice, the geometry of the dice, etc.) on could in principle predict the result of a throw of dice, and the phenomenon would loose its probabilistic character. However, it is quite obvious and intuitive that the outcome of different experiments on dice would be highly sensitive to to the initial conditions and that it would require an enormous amount of information to make the prediction. It is therefore much more efficient in practice to perform a probabilistic description of the problem, where one imposes some ignorance on the initial conditions which are said to be chosen "at random". This phenomenon is encountered in celestial mechanics, and many other problems, when initial conditions are close but not "infinitesimally" close, provided the time of evolution is long enough.

The case of three unequal mass planets orbiting around a "sun" taking into account their mutual interactions is shown on Fig. (4.2). At the beginning, everything evolves rather smoothly. However, after some time, the lightest planet is simply ejected from the system; this is of course compatible with energy conservation, which would not be the case for a two-body system. By letting the computer run for a longer time, the two other planets, which have a smooth motion at first, also reach unexpected configurations.

---

[11] Jacques Laskar, *Is the Solar System stable ?*, Progress in Mathematical Physics, 66, pp 239–270, (2013); G.J. Sussman, J. Wisdom; *Chaotic evolution of the solar system*, Science 257 (1992), 56–62.

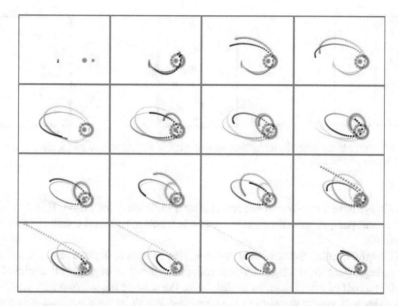

Fig. 4.2 Evolution of three planets around a star taking into account their mutual interactions. The time sequence of the pictures must be read from left to right and from bottom to top. The time interval between two pictures is the same. One sees that at the eleventh stage, the third planet, lighter and initially close to the second one, is expelled from the system. Pictures due to Jean-François Colonna, colonna@cmap.polytechnique.fr, http://www.lactamme.polytechnique.fr; all rights reserved

4.6.3 Poincaré's Recurrence Theorem

Poincarè's recurrence theorem[12] (1890) says that for almost all "initial conditions", a dynamical conservative system with a phase space of *finite volume* will come back with time as close as you want to its initial condition, and this repeatedly. That the volume of the phase space is bounded is not exceptional: if the energy E is constant, and the potential V is positive, the positivity of the kinetic energy ensures that the phase space is bounded by the condition $V(\mathbf{r}) \leq E$.

To establish the theorem, consider a point P point in a D_0 neighborhood of the phase space. Of course, it will evolve and find itself, at time t in a different region D_1 of the phase space. These two regions may not have anything in common, but they have the same volume, according to the Liouville theorem. After kt, with k as big as you want, the system will end up in a D_k neighborhood, always of the same volume. It can browse through a region with as many D_k domains as you want. But, since the volume of these domains is the same and the total volume of the space of the phases is finite, while their number is as large as we want, necessarily two of these domains

---

[12] Barreira, Luis (2006). Zambrini, Jean-Claude (ed.). *Poincaré recurrence: Old and new.* XIVth International Congress on Mathematical Physics. World Scientific. pp. 415422. doi:10.1142/9789812704016_0039.

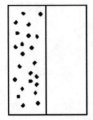 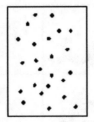 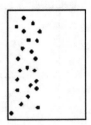

Fig. 4.3 Zermelo paradox: the molecules, all to the left of the chamber at $t = 0$, are found there again after a certain time T

will always have a non-zero intersection after a sufficiently long time, This dynamic system will pass, at some time t as close as one wants to its initial condition, and this repeatedly.

This effect, called the "Zermelo paradox", thus seems to deprive us of the second principle of thermodynamics. Indeed, imagine molecules of a gas all enclosed at $t = 0$ in a half of a container, as in Fig. (4.3). We remove the intermediate wall, the molecules will spread throughout the container. But Poincaré's theorem says there is a time T where they all end up in the initial half!

The paradox of Zermelo,[13] is schematized on Fig. 4.3. Of course, this time T is quite long. It is estimated to be of the order of $2^{10^{23}}$ times a characteristic time of molecule migration, which easily leads to some incomparably greater duration than any extrapolation of the age of the universe.

4.6.4 The Butterfly Effect; the Lorenz Attractor

The most famous strange attractor is probably the Lorenz attractor, after its inventor Edward N. Lorenz,[14] who discovered it in 1963 starting from a mathematical model of the atmosphere. This gave a spectacular new interest in chaos.

Consider the evolution of a rectangular slice of the atmosphere which is heated from below and cooled from above. There are three variables, x, the convective flow of the atmosphere, y the horizontal temperature distribution and z the vertical temperature distribution.

The details of the physics involved is of little interest here. In the Lorenz model, the evolution of these variables is given by the (Hamiltonian) non-linear differential system

[13] Ernst Zermelo, *Über einen Satz der Dynamik und die mechanische Wärmetheorie*, Wied. Ann. 57 (1896, 793).

[14] Edward N. Lorenz, *Deterministic Nonperiodic Flow*, Journal of the Atmospheric Sciences, vol. 20, n⁰2, march 1963, p. 130–141; Lorenz, Edward N., *The essence of chaos*, University of Washington Press, 1993.

$$\frac{dx}{dt} = \sigma(y - x)$$

$$\frac{dy}{dt} = \rho x - y - xz$$

$$\frac{dz}{dt} = xy - \beta z, \tag{4.41}$$

where σ is the ratio between the viscosity and the thermal conductivity, ρ is the temperature difference between the top and the bottom of the slice, and β is the ratio of the width and the height of the slice.

Lorenz used to solve this problem numerically using hours of computer time at night,[15] by standard successive iteration techniques $(x_i, y_i, z_i) \rightarrow (x_{i+1}, y_{i+1}, z_{i+1})$. At that time, this generated kilograms of paper called computer listings. One day, Lorenz had the idea of redoing a calculation whose solution he had found the day before, using as a starting point not the last point obtained the day before, but some intermediate value (x_i, y_i, z_i) obtained in the calculation. To his great surprise, after a relatively small number of iterations, the following values appeared completely different from those obtained previously. Lorenz had rediscovered chaos, due, in that case, to round-off errors of the numbers he used.

The sensitivity of the results to the initial conditions induces the same type of difference between two solutions initially close to one another. Lorenz called that the "butterfly effect". Actually, the title of one of his talks was: *Can the beat of a butterfly's wing in Brasil cause a tornado in Texas?* Whether or not it is a coincidence, the "Lorenz attractor" has the shape of the wings of a butterfly.

On Figs. (4.4) and (4.5) one can see the result of an iteration of the Eq. (4.41). We notice that the time evolution of the point (x, y, z) has a perfectly quiet behavior: the point turns around on a wing of the attractor, but that unexpectedly it "jumps"

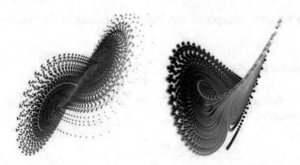

Fig. 4.4 Lorenz attractor viewed from two different sides. The points correspond to a discrete numerical iteration of (4.41). One can follow the points and observe the sudden and unexpected transition from one wing of the attractor to the other, which was not possible to predict half a semi-period before. (Courtesy Jean-François Colonna)

---

[15] As an order of magnitude of the performances of computers in the early 1960's, his first computer, called Royal McBee, could perform 60 multiplications per second.

Fig. 4.5 Projection of the
Lorenz attractor on the (x,z)
plane. (Courtesy
Jean-François Colonna)

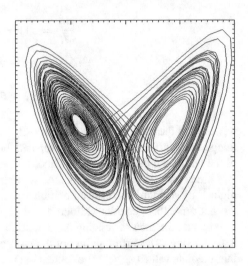

from one wing to the other at certain times. This occurs unexpectedly in space as
well as in time, in the sense that the trajectories of two points which are initially very
close (in positions and velocities) can become completely different at a later time.
In particular, the two positions can be on different wings of the attractor.

The rigorous proof was only given in 2001 by Warwick Tucker.[16]

4.7 Exercises

4.1. Poisson brackets for the angular momentum

Calculate the Poisson brackets of the three components of the angular momentum
$\mathbf{L} = \mathbf{r} \times \mathbf{p}$.

4.2. The Lenz vector in a two-body $1/r$ potential.

We consider Kepler's problem $H = p^2/2m - e^2/r$. Calculate the Poisson brackets
of the components of the Lenz vector

$$\mathbf{A} = \frac{\mathbf{p} \times \mathbf{L}}{m} - e^2 \frac{\mathbf{r}}{r}$$

between each other, with the components of the angular momentum and with the
Hamiltonian. What can one conclude on the number of unknowns in that problem?

---

[16] W. Tucker, *A Rigorous ODE Solver and Smale's 14th Problem*, Found. Comp. Math., vol. 2,
2002, p. 53–117.

4.3. Three coupled oscillators

The problem of three coupled oscillators is treated in analogous manner as the two-body case of Sect. (4.3) with the Jacobi variables.
The Hamiltonian is

$$H = \frac{1}{2m}(p_1^2 + p_2^2 + p_3^2) + \frac{m\omega^2}{2}(x_1^2 + x_2^2 + x_3^2)$$
$$+ \frac{m\Omega^2}{2}((x_1 - x_2)^2 + (x_2 - x_3)^2 + (x_3 - x_1)^2).$$

The canonical transformation (Jacobi variables) is

$$X_1 = (x_1 - x_2)/\sqrt{2}, \quad X_2 = (2x_3 - x_1 - x_2)/\sqrt{6}, \quad X_3 = (x_1 + x_2 + x_3)/\sqrt{3}$$
$$P_1 = (p_1 - p_2)/\sqrt{2}, \quad P_2 = (2p_3 - p_1 - p_2)/\sqrt{6}, \quad P_3 = (p_1 + p_2 + p_3)/\sqrt{3}$$

which gives: $\sum_i P_i^2 = \sum_i p_i^2$, and $3(X_1^2 + X_2^2) = (x_1 - x_2)^2 + (x_2 - x_3)^2 + (x_3 - x_1)^2$.

Calculate The poisson brackets $\{X_i, P_j\}$.

Describe the three-body trajectory (or manifold) in phase space.

4.4. Forced oscillations

Consider a one dimensional harmonic oscillator of Hamiltonian

$$H = \frac{p^2}{2m} + \frac{1}{2}m\omega^2 x^2 \tag{4.42}$$

where x and p are Lagrange conjugate variables.

1. We set $x = X/\sqrt{m\omega}$ and $p = P\sqrt{m\omega}$.
   Write the expression of the Hamiltonian in terms of X and P, and calculate the Poisson bracket $\{X, P\}$.
2. We introduce the functions a and a^*, the complex conjugate of a, defined by

$$a = \frac{X + iP}{\sqrt{2}}, \qquad a^* = \frac{X - iP}{\sqrt{2}}.$$

   Write the Hamiltonian in terms of a, a^* and ω.
3. Calculate the Poisson bracket $\{a, a^*\}$.
4. Write the time evolution equation of a and give its general solution.
   Write the energy E of the oscillator in terms of the parameters of this solution and of ω.

5. We assume the energy of the oscillator is zero for $t \leq 0$, $E(t \leq 0) = 0$. Between $t = 0$ and $t = T$ one applies to the oscillator a force which derives from the potential energy $H_{pot} = b\sqrt{2}X \sin(\Omega t)$ ($H_{pot} = 0$ if $t \leq 0$ or $t > T$) where b is a parameter. Calculate the energy E' of the oscillator for $t > T$.
6. Discuss the variation of E' as a function of the exciting frequency Ω.

4.8 Problem. Closed Chain of Coupled Oscillators

We recall that for $1 \leq n \leq N$ and $1 \leq n' \leq N$

$$\frac{1}{N} \sum_{k=1}^{N} \exp\left(\frac{2ik(n-n')\pi}{N}\right) = \delta_{nn'} \quad \text{(Kronecker } \delta\text{)}.$$

We consider a closed chain of N particles of equal mass m placed on a plane circle (see Fig. 4.6). These particles have each a one dimensional motion along the direction (x) perpendicular to the plane. We note x_n, $n = 1, \ldots, N$ the abscissa of particle n along this axis.

These particles form a set of harmonic oscillators coupled to their nearest neighbors. The Hamiltonian is

$$H = \sum_{n=1}^{N} \left[\frac{p_n^2}{2m} + \frac{1}{2}m\omega^2 x_n^2 + \frac{1}{2}m\Omega^2(x_n - x_{n+1})^2\right] \quad (4.43)$$

Fig. 4.6 Chain of coupled oscillators

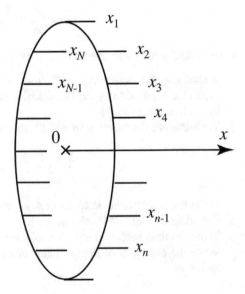

where p_n is the conjugate momentum to x_n and where we use the cyclic convention $x_{N+1} \equiv x_1$.

1. Define the following complex variables

$$y_k = \frac{1}{\sqrt{N}} \sum_{n=1}^{N} e^{2ikn\pi/N} x_n \,, \quad q_k = \frac{1}{\sqrt{N}} \sum_{n=1}^{N} e^{-2ikn\pi/N} p_n \qquad (4.44)$$

whose inverse relations are

$$x_n = \frac{1}{\sqrt{N}} \sum_{k=1}^{N} e^{-2ikn\pi/N} y_k \,, \quad p_n = \frac{1}{\sqrt{N}} \sum_{k=1}^{N} e^{2ikn\pi/N} q_k. \qquad (4.45)$$

(a) Show that

$$y_k = y_{N-k}^* \,, \quad q_k = q_{N-k}^*.$$

(b) Show that

$$\sum_{k=1}^{N} y_k y_k^* = \sum_{n=1}^{N} x_n^2 \quad \text{and} \quad \sum_{k=1}^{N} q_k q_k^* = \sum_{n=1}^{N} p_n^2. \qquad (4.46)$$

(c) Show that

$$\sum_{n=1}^{N} (x_n - x_{n+1})^2 = \sum_{k=1}^{N} 4 \sin^2\left(\frac{k\pi}{N}\right) y_k y_k^*. \qquad (4.47)$$

2. *Equations of motion and their solutions*

   (a) Write the Hamiltonian (4.43) in terms of the variables $\{y_k, y_k^*, q_k, q_k^*\}$.
   (b) Calculate the following Poisson brackets

$$\{y_j, q_k\} \,, \quad \{y_j^*, q_k^*\} \,, \quad \{y_j, q_{N-k}^*\} \,, \quad \{y_j^*, q_{N-k}\}. \qquad (4.48)$$

   (c) Write the differential equations satisfied by the variables $\{y_k, y_k^*, q_k, q_k^*\}$.
   (d) Write the general expression of $\{y_k(t)\}$; deduce from it the expression of $\{x_n(t)\}$.

3. We assume that at time $t = 0$ we have $y_N(0) = 1$, $\dot{y}_N(0) = 0$ and $\{y_n(0) = 0, \dot{y}_n(0) = 0, \forall n \neq N\}$. Calculate $\{x_n(t)\}$ and interpret the result.

4. *Propagation of waves.*
   We now assume, for simplicity, that $\omega = 0$. We also assume that $N \gg 1$, so that $\sin(k\pi/N) \simeq (k\pi/N)$ for $k \ll N$. We assume that for $t = 0$ we have $y_{N-1} = 1$, $y_1 = 1$, $y_n = 0$ if $n \neq (1 \text{ or } N - 1)$, and $\dot{y}_n = 0 \ \forall n$.

(a) Calculate $x_n(t)$ and $x_{N-n}(t)$.
(b) Interpret the result physically.
(c) We assume that the distance between two neighboring oscillators is a. Setting that $x_n(t)$ is the value of a function $f(t, y)$ for $y = na$, write the wave equation (second order partial differential equation) satisfied by the function f.

Chapter 5
Action, Optics, Hamilton-Jacobi Equation

William Rowland Hamilton (1805–1865) played a leading role in theoretical physics and mathematics in the 19th century—he was famous, among other things, for his discovery of quaternions on October 16, 1843, which he inscribed on the *Broome bridge* in Dublin with the formulas $i^2 = j^2 = k2 = ijk = -1$. Born in Dublin, a child prodigy, he knew thirteen languages, including Arabic, Sanskrit and Persian, at the age of thirteen. At sixteen he mastered Newtons *Principia* and other contemporary works of differential calculus, and the following year he immersed himself in Lagranges texts, and in Laplace's *Celestial Mechanics*. In particular, he discovered an error in a demonstration of Laplace, which made him noticed by confirmed scholars in Dublin. He won several gold medals in various competitions, but even before completing his university degree, the scientific community pushed him to be appointed, at the age of twenty-two, to the position of *Royal Astronomer* of Ireland, then to Professor of Astronomy at Trinity College, Dublin University. He moved to the observatory at Dunsink, near Dublin. He knew theoretical astronomy, but absolutely nothing in practical astronomy. Nevertheless, he was interested in the development of optical instruments and, in particular, the question of *caustics*, (Fig. 5.1) which limited the precision of the images, and thereby the resolution of optical instruments.

In 1824, at the age of 19, Hamilton presented an article at the Royal Irish Academy entitled *Caustics*. The Reading Committee recognized its originality and value, but recommended that the article be developed and simplified before publication. This will be the starting point of an incredible series of publications, which became, between 1825 and 1828, a true treatise not only of optics but of analytical mechanics!

Between 1825 and 1828, the article grew considerably, mainly through the addition of details requested by the committee; but it was also made much more understandable as the peculiarities of this methodology became perfectly apparent to his contemporaries.

In 1827 he introduced the theory of a single function, the *characteristic function*, which united mechanics, optics, and mathematics and helped to establish the wave

© The Author(s), under exclusive license to Springer Nature Switzerland AG 2023   87
J.-L. Basdevant, *Variational Principles in Physics*,
https://doi.org/10.1007/978-3-031-21692-3_5

Fig. 5.1 Examples of caustics: **a** inside a ring; **b** light coming out of a glass of water; **c** caustic of a parallel beam reflected by an ideal cylindrical mirror

theory of light. His three main articles were eventually named *Theory of Systems of Rays* and published from 1828 in *Transactions of the Royal Irish Academy*.[1] To this must be added his three fundamental publications on mechanics.[2]

Hamilton was fascinated by the *action* and the Fermat principle. He understood that it is a principle of *stationary* action and not minimal (in a concave mirror, there are two images of the same object and not just one). The variational principle, also known as the Hamilton principle, is the essential element of these articles. This principle, reformulated by Jacobi, resulted in a formulation different from classical mechanics; it is currently known as Hamiltonian mechanics.

This formulation, like the Lagrangian mechanics on which it is based, is, at first glance, very mathematical without new impact on physics, it constitutes a more powerful method for solving complex problems. Lagrangian and Hamiltonian mechanics were developed to describe the movement of discrete systems; they were extended to continuous systems and field theory.

In Sect. 5.1 we will briefly describe Hamiltons results on geometric optics, where he chooses to work directly with *the action* in the form of his *characteristic function S*, function of canonical variables (x, p) and not with the Lagrangian or the Hamiltonian.

Then we'll see how Hamilton formalized the fact that geometrical optics is the limit of wave optics for short wavelengths, and we will see its amazing structural similarity with mechanics (which Hamilton discovered a hundred years before the discovery of quantum mechanics). In the approximation of small wavelengths known as the eikonal approximation, the wave propagates with a wave vector locally perpendicular to *geometric wave fronts*. This corresponds exactly to the Fermat principle, and the geometric interpretation is nothing else than the *Huygens-Fresnel principle*.

---

[1] *Theory of systems of rays*, Transactions of the Royal Irish Academy, vol. 15, 69, 1828; *Second supplement*, ibid. vol. 16, 93, 1831; *Third supplement*, ibid. vol. 17, 1, 1837.

[2] *On a General Method of expressing the Paths of Light and of the Planets by the Coefficients of a Characteristic Function*, Dublin University Review, 795–826 (1833); *On the Application to Dynamics of a General Mathematical Method previously applied to Optics*, British Association Report, pp. 513–518, (1835); *On a General Method in Dynamics*, Philosophical Transactions of the Royal Society, pp. 247–308, (1834).

In Sect. 5.2 we will explain the formalism of Hamilton's characteristic function, in other words the use of the action itself as a physical quantity in the equations of motion of mechanics. This will lead us to the Hamilton-Jacobi equation. We will discover a series of important results. We shall see how, for conservative systems, the *flow* of trajectories is orthogonal to the surfaces of constant action. This will make us rediscover the Maupertuis principle in a geometric form perfectly analogous to the Fermat principle, it will again show the similarity of classical mechanics and geometrical optics, as well as what results in quantum mechanics, which we will see in Sect. (5.3).

In Sect. (5.4), we will end by showing on a few examples the many applications of the *Hamilton-Jacobi equation*, first order non-linear equation.

5.1   Geometrical Optics, Characteristic Function of Hamilton

Geometrical Limit of Wave Optics

Hamilton seeks to show how geometrical optics presents itself as the limit at small wavelengths of wave optics. In particular, he wants to show the Fermat principle in a wave propagation equation.   For simplicity, we will not follow the original wording of Hamilton.[3] To simplify the presentation, we will consider the propagation of a scalar wave (and not vector as in the Maxwell equations[4]) in a variable refractive index medium. The general case of the propagation of electromagnetic waves in an anisotropic environment, non-conductive of electrical and magnetic susceptibilities ε and μ, taking into account possible discontinuities between two media and polarization, is treated in the book of Born and Wolf *Principles of Optics* [14], Chap. 3 and Appendix I. It is enough for our purpose to consider a non-magnetic isotropic medium, and the result is basically identical to the evolution of a scalar wave.

Consider the propagation of a scalar wave Φ in a variable refractive index medium $n(\mathbf{r})$, assuming that the medium is inhomogeneous, but isotropic: the refractive index n depends on the point considered but not on the direction of propagation.

The propagation equation of this wave $\Phi(\mathbf{r}, t)$ is

$$\frac{n^2}{c^2} \frac{\partial^2 \Phi}{\partial t^2} - \nabla^2 \Phi = 0. \tag{5.1}$$

We consider a periodic wave of pulse ω, that is $\Phi(\mathbf{r}, t) = \varphi(\mathbf{r})e^{-i\omega t}$, which inserted in the previous equation, leads to

---

[3] *Theory of systems of rays*, Transactions of the Royal Irish Academy, vol. 15, 69, 1828; *Second supplement*, ibid. tbf 16, 93, 1831; *Third supplement*, ibid. vol 17, 1, 1837.

[4] The validity of the results was demonstrated in 1911 by A. Sommerfeld and J. Runge *Ann. d. Physik, 35, 289 (1911)*.

$$\frac{n^2 \omega^2}{c^2}\varphi + \nabla^2\varphi = 0. \tag{5.2}$$

We seek a solution to this equation of the form

$$\varphi = \varphi_0(\mathbf{r})e^{ik_0 S(\mathbf{r})}, \tag{5.3}$$

where

$$k_0 = \omega/c = (2/\pi)/\lambda_0, \tag{5.4}$$

is the modulus of the mean wave vector in the vicinity of the relevant $\mathbf{r}$ point, λ_0 being the corresponding wavelength.

Hamilton calls the quantity S in (5.3) the *characteristic function*. In 1895, in an independent study, Burns gave it the name of *eikonal* (from the Greek $\varepsilon\iota\kappa\omega\nu$; image or picture), name that was later kept for Hamilton's function.

If we insert (5.3) in (5.2), we get, after simplification by $e^{ik_0 S(\mathbf{r})}$ and dividing by k_0^2,

$$\varphi_0(\nabla S)^2 - \frac{i}{k_0}\left(2\nabla\varphi_0 \cdot \nabla S + \varphi_0\nabla^2 S\right) - \frac{1}{k_0^2}\nabla^2\varphi_0 = n^2\varphi_0. \tag{5.5}$$

In this equation, the cancellation of the imaginary term proportional to $1/k_0$ can be written, multiplying by φ_0,

$$\nabla \cdot (\varphi_0^2 \nabla S) = 0. \tag{5.6}$$

This is a conservation equation, in this case conservation of energy. The wave propagates in the direction of ∇S and the energy density is proportional to φ_0^2 (Fig. 5.2). One will find the complete interpretation in terms of the Poynting vector in Born and Wolf's book [14], Chap. 3.

Consider the real part, reinstating the wavelength λ,

$$\varphi_0|\nabla S|^2 - \frac{\lambda^2}{4\pi^2}\nabla^2\varphi_0 = n^2\varphi_0. \tag{5.7}$$

Fig. 5.2 Local direction of a light beam with respect to a surface area $S = $ constant in the three-dimensional space

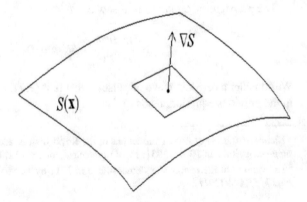

Suppose that the wavelength is very small, that is the n index does not vary over a wavelength and that the size of instruments (for example diaphragms) is much larger than λ defined in (5.4). This hypothesis can also be expressed as $\lambda \to 0$ therefore $k_0 \to \infty$. For Hamilton, this is the limit of *geometrical optics*.

We then neglect the term in $1/k_0^2$ which leads to the *fundamental equation of geometrical optics*

$$(\nabla S)^2 = n^2 \tag{5.8}$$

nowadays called the eikonal equation.

In other words, the phenomena of interference and diffraction are found in the term $\lambda^2/(4\pi^2)\nabla^2\varphi_0$ and in the boundary conditions (edges, diaphragms).

In this approximation, the surfaces $S = C^{st}$ are the *geometric wave fronts* of the wave

$$\Phi(\mathbf{r}, t) = \varphi_0(\mathbf{r})e^{i(k_0 S(\mathbf{r})-\omega t)}, \tag{5.9}$$

which propagates with a $\mathbf{k}$ wave vector *locally perpendicular* to the geometrical wave fronts, on which the S function is constant $S(\mathbf{r}) = C^{st}$.

This situation is illustrated in Fig. 5.3.

This orthogonality in $\mathcal{R}^3$ is quite fundamental.

If $\mathbf{r}(s)$ is the position of a point P of a light ray, at the arc length s of that point P on the radius, then $d\mathbf{r}/ds = \mathbf{s}$ is a unit vector at that point, and the radius equation is $n(\mathbf{r})(d\mathbf{r}/ds) = \nabla S(\mathbf{r})$. Between two wavefronts S and $S + dS$ we therefore have:

$$n\frac{d\mathbf{r}}{ds} = \nabla S \quad \text{and} \quad \frac{dS}{ds} = \frac{d\mathbf{r}}{ds} \cdot \nabla S = n. \tag{5.10}$$

The geometric interpretation of (5.2) or (5.10), is nothing but the *Huygens-Fresnel principle*. This principle, first wave theory of light, consists in saying that the light propagates like a wavefront. At every moment t, each point of the wavefront can be

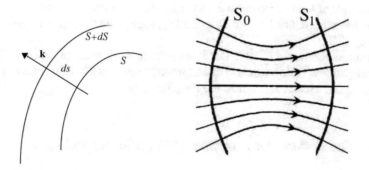

Fig. 5.3 a Two surfaces of S = constant, of infinitesimal separation ds where the light ray $\mathbf{k}$, perpendicular to the surfaces, follows a path ds of optical length nds; **b** Flow of rays in the vicinity of a divergent device

Fig. 5.4 Fermat Principle

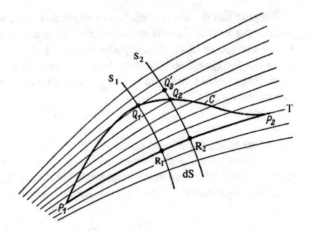

considered as a point source. At the next moment $t + \delta t$ the new wave front is the envelope of the light ray spheres $\delta r = (c/n)\delta t$ centered at each point of the previous wave front (Fig. 5.4).

As for the Fermat principle, the optical path λ on a C curve with a n index is $\lambda = \int_C n \, ds$, the corresponding light travel time being $\tau = n\lambda/c$.

Consider, in a beam, the light ray T resulting from P_1 passing through a point R_1 of the surface area S_1 and ending in P_2 by passing through R_2 on S_2. Any other path from P_1 to P_2 that does not pass along the $R_1 R_2$ radius would, by triangular inequality, be longer. S_1 and S_2 being at equal distance, the $Q_1 Q_2$ portion is greater than $Q_1 Q_2'$ which is equal to $R_1 R_2$. The radius follows the shortest optical path as required by the Fermat principle: $\delta \int n(\mathbf{r}) ds = 0$. Hamiltons characteristic function principle is equivalent to the Fermat principle within the bounds of the eikonal approximation.

5.2   Action and the Hamilton-Jacobi Equation

The principle of least action consists in finding the equations of the motion by minimizing the action defined according to the Lagrangian and the departure and arrival points by (3.2).

Hamilton discovered in 1831 that it is useful to work with the action itself as a physical quantity. To do so, we will first express the action as a function of coordinates and time $S(x_1, x_2, \ldots, x_n; t)$, then use the properties we know.

5.2.1   *The Action as a Function of Coordinates and Time*

For one degree of freedom, the question is to calculate the values of S along the set of physical trajectories, that is, as a function of the point and time of arrival (x, t), The starting point and time being fixed.

Generally speaking, we will characterize the various trajectories from (x_1, t_1) and arriving at (x, t) by the value of the action $S(x, t; x_1, t_1)$.

The action is defined by

$$S = \int_{t_1}^{t} \mathcal{L}(x, \dot{x}, t') \, dt', \tag{5.11}$$

the $(x(t), \dot{x}(t))$ variables assuming in this expression their physical values, which satisfy the Lagrange-Euler equations.

The variation of the action written in (3.5) is

$$\delta S = \int_{t_1}^{t} \left(\frac{\partial \mathcal{L}}{\partial x} \delta x(t) + \frac{\partial \mathcal{L}}{\partial \dot{x}} \delta \dot{x}(t) \right) dt. \tag{5.12}$$

We integrate the second term by parts, but we do not impose any more to arrive at the same point $x(t)$ but in a neighbouring point $x(t) + \delta x(t)$ (by maintaining $\delta x(t_1) = 0$) the integrated term therefore no longer disappears, and we obtain

$$\delta S = \frac{\partial \mathcal{L}}{\partial \dot{x}} \delta x(t) + \int_{t_1}^{t} \left(\frac{\partial \mathcal{L}}{\partial x} - \frac{d}{dt} \left(\frac{\partial \mathcal{L}}{\partial \dot{x}} \right) \right) \delta x(t) \, dt. \tag{5.13}$$

By hypothesis, the trajectory is physical, and the right hand side integral cancels.

We thus obtain a variation of the action

$$\delta S = \frac{\partial \mathcal{L}}{\partial \dot{x}} \delta x(t) = p \, \delta x(t), \tag{5.14}$$

or, in general,

$$\delta S = \sum_{i=1}^{N} p_i \, \delta x_i. \tag{5.15}$$

if we work with a $[q_i, p_i]$ set of *canonically conjugate variables*.

Time Dependence

Similarly, if we change the arrival moment t, the action being the integral of the lagrangian (5.11), we have, obviously,

$$\frac{dS}{dt} = \mathcal{L}. \tag{5.16}$$

If the action is seen as a function of coordinates and time, we have

$$\frac{dS}{dt} = \frac{\partial S}{\partial t} + \sum_{i=1}^{N} \frac{\partial S}{\partial x_i} \dot{x}_i = \frac{\partial S}{\partial t} + \sum_{i=1}^{N} p_i \, \dot{x}_i. \tag{5.17}$$

By combining these two equalities, we obtain that the partial derivative of the action with respect to time is, up to its sign, equal to the *Hamiltonian*

$$\frac{\partial S}{\partial t} = \mathcal{L} - \sum_{i=1}^{N} p_i \, \dot{x}_i = -H, \tag{5.18}$$

and the total differential of the action is therefore written in terms of coordinates and time

$$dS = \sum_{i=1}^{N} p_i \, dx_i - H \, dt. \tag{5.19}$$

Since no reference is made to the point and time of departure, the formal expression of the action is therefore

$$\delta S = \delta \int \left(\sum_{i=1}^{N} p_i \frac{dx_i}{dt} - H \right) dt \equiv \delta \int \mathcal{L} \, dt = 0, \tag{5.20}$$

5.2.2   Least Action Principle

Hamilton's least action principle is written $\delta S = 0$, indeed Eq. (5.20) simply gives

$$\delta S = \delta \int \left(\sum_{i=1}^{N} p_i \frac{dx_i}{dt} - H \right) dt \equiv \delta \int \mathcal{L} \, dt = 0, \tag{5.21}$$

which is the form (3.2) that served us as *starting point* in Chap. 3.

However, in this case we work with *conjugate variables* (x, p) and not the $(x, \dot{x})$ variables as in Chap. 3!

Hamilton's canonical equations are directly deduced of the expression (5.20) of the action. Indeed, consider the variables x and p as independent. In the case of a single degree of freedom, the action is

$$S = \int_{(1)}^{(2)} (p \, dx - H \, dt). \tag{5.22}$$

Lets vary x by δx and p by δp by imposing $\delta x(2) = \delta x(1) = 0$. The variation of S is

$$\delta S = \int_{(1)}^{(2)} \left(\delta p\, dx + p\, d(\delta x) - \frac{\partial H}{\partial x} \delta x\, dt - \frac{\partial H}{\partial p} \delta p\, dt \right). \qquad (5.23)$$

The second term in the integral can be integrated by parts. The integrated term $(p, \delta x)$ cancels out since by assumption $\delta x(2) = \delta x(1) = 0$, and we get

$$\delta S = \int_{(1)}^{(2)} \left(\delta p\, [dx - \frac{\partial H}{\partial p}\, dt] - \delta x\, [dp + \frac{\partial H}{\partial x}\, dt] \right), \qquad (5.24)$$

which cancels for any variation $(\delta x, \delta p)$ if and only if the terms to be integrated are identically zero that is to say

$$dx - \frac{\partial H}{\partial p}, dt = 0, \quad dp + \frac{\partial H}{\partial x}, dt = 0,$$

which are simply the canonical equations of Hamilton.

5.2.3 Hamilton-Jacobi Equation

From the expressions (5.18) and (5.15), we can replace in the Hamilton function the conjugate momenta p_i by the partial derivatives of the action. This leads to the *Hamilton-Jacobi equation*

$$\frac{\partial S}{\partial t} + H(x_1, \ldots, x_N, \frac{\partial S}{\partial x_1}, \ldots, \frac{\partial S}{\partial x_N}; t) = 0. \qquad (5.25)$$

This equation is a *single* non-linear partial differential equation of first order. In the same way as the equations of Lagrange-Euler or as the canonical equations, it allows to calculate the motion.

The function $S(\{x_i\}; t)$ completely determines the motion, it is called the Hamilton *main function*.

The use of either of these formalisms is a matter of convenience or mathematical structure of the problem.

The Hamilton-Jacobi equation is particularly suitable for the separation of variables and the choice of appropriate variables in the problem, to the symmetry of a problem, as we will see below. One can refer to the book by Landau and Lifshitz, [1] *Mechanics* Chap. 7, Sect. 48, for a discussion of this point, which involves a formulation of canonical transformations more complete than what we have done in Chap. 3 of the previous chapter.

5.2.4 Conservative Systems, Reduced Action, Maupertuis Principle

Reduced Action

Suppose Hamiltons H function does not dependent explicitly on time. Then the energy is conserved. Let E be the energy value of the problem under consideration, the Eq. (5.18) translates as

$$\frac{\partial S}{\partial t} = -E \quad , \tag{5.26}$$

that is

$$S = -Et + S_0(x_1, \ldots, x_N) \quad , \tag{5.27}$$

The quantity S_0 is called the *reduced action*. It satisfies the equation

$$H(x_1, \ldots, x_N, \frac{\partial S_0}{\partial x_1}, \ldots, \frac{\partial S_0}{\partial x_N}) = E. \tag{5.28}$$

In general, referring to (5.20) we define the reduced action S_0 by

$$S_0 = \int \left(\sum_{i=1}^{N} p_i \, dx_i \right), \tag{5.29}$$

and, for a conservative system, we see that the variational principle applies to this quantity: $\delta S_0 = 0$.

Geometrical Interpretation

The relation (5.15) can also be written in terms of the reduced action

$$\frac{\partial S_0}{\partial x_i} = p_i. \tag{5.30}$$

This form shows a simple geometric property that makes a direct link with optics. Lets put ourselves in Cartesian coordinates for clarity and consider the simple case where the momenta merge with motion quantities $p_i = m_i \dot{x}_i$. Consider in the space the coordinates $(x_1, x_2, \ldots, x_N)$, surfaces on which the reduced action is constant $S_0 = C^{st}$. The relationship (5.30) means that the vector $P \equiv (p_1, p_2, \ldots, p_N)$ is at all points orthogonal to these surfaces (Fig. 5.5).

If we consider the simple case of a particle in 3-dimensional space, we see that the trajectory is, at any point of space, orthogonal to the surface $S_0 = C^{st}$ passing by this point. At a given time, this property is also valid for the action S. If we note $d\tilde{r}$

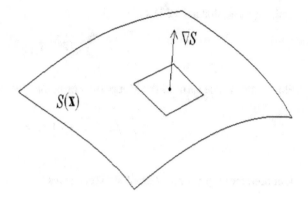

Fig. 5.5 Trajectory and orthogonality of the surface S_0=constant if the energy is conserved. This is a co-dimension 1 surface in configuration space

an elementary vector tangent to the area $S_0 = C^{st}$ at point $\mathbf{r}$, we have by definition $\nabla S_0 \cdot d\tilde{\mathbf{r}} = 0$. Hence

$$\nabla S_0 \cdot d\tilde{\mathbf{r}} = \mathbf{p} \cdot d\tilde{\mathbf{r}} = 0. \tag{5.31}$$

In other words, the *trajectory flow is orthogonal to surfaces* $S_0 = C^{st}$.

The similarity between mechanics and geometrical optics, light rays and characteristic function, is obvious. This was Hamiltons discovery!

Maupertuis Principle

For a mass particle m in a potential $V(\mathbf{r})$, the Eq. (5.28) is written

$$\frac{1}{2m}(\nabla S_0)^2 + V(\mathbf{r}) = E \quad , \quad \text{or} \quad (\nabla S_0)^2 = 2m(E - V(\mathbf{r})). \tag{5.32}$$

In this problem, the momentum is simply equal to the linear $\mathbf{p} = m\dot{\mathbf{r}}$. The reduced action (5.29) is therefore written

$$S_0 = \int \mathbf{p} \cdot d\mathbf{r} = \int m\,\dot{\mathbf{r}} \cdot d\mathbf{r}. \tag{5.33}$$

In this expression, we want to express everything in terms of the position variable. If we designate by ℓ the curvilinear abscissa following the trajectory $\mathbf{r}(t)$, we obviously have

$$(d\mathbf{r})^2 \equiv (dx^2 + dy^2 + dz^2) = d\ell^2, \quad \text{hence} \quad \dot{\mathbf{r}} \cdot d\mathbf{r} = \dot{\ell}\,d\ell. \tag{5.34}$$

But the kinetic energy is $T = m\dot{\mathbf{r}}^2/2 = m\dot{\ell}^2/2$, and since $T = m\dot{\mathbf{r}}^2/2 = E - V$, we obviously get

$$\dot{\ell} = \sqrt{\frac{2(E - V)}{m}}, \tag{5.35}$$

and, inserting this in (5.34) and (5.33),

$$S_0 = \int \sqrt{2m(E - V)} \, d\ell. \tag{5.36}$$

Hence the simple form of the Maupertuis principle given in Chap. 2 Sect. (2.3).

$$\delta \int \sqrt{2m(E - V)} \, d\ell = 0. \tag{5.37}$$

Geometrical Optics and Classical Mechanics

Of course, we note the great similarity of the equation of the eikonal (5.8) and the Hamilton-Jacobi equation (5.32) for a material point. The S_0 reduced action of the latter and the eikonal S for a light wave follow the same law if we do the correspondence

$$n(\mathbf{r}) \Longleftrightarrow \sqrt{2m(E - V(\mathbf{r}))}. \tag{5.38}$$

One just has to go backwards on the path that leads to to (5.32), in particular (5.21), to see that the eikonal approximation corresponds exactly to the Fermat principle

$$\delta \int n(\mathbf{r}) d\ell = 0 \Longleftrightarrow \delta T = \delta \int \frac{n(\mathbf{r})}{c} d\ell = 0. \tag{5.39}$$

The Fermat Principle and the Maupertuis Principle (5.37) have an obvious similarity if one makes the correspondence (5.38).

Hamilton made this discovery in 1834. He had understood, in 1830, how and in what limits the geometrical optics was an *approximation* of wave optics. Fascinated by variational principles, and in particular by the similarity between the Maupertuis principle in mechanics and the Fermat principle in geometric optics, he made in 1830 the surprising remark that the formalisms of optics and could be unified, and (prophetic vision) that the Newtonian mechanics corresponds to the same *limit* or approximation, as geometrical optics compared to wave optics.

This remark was ignored by his contemporaries what deplored in 1891 the famous mathematician Felix Klein (1849–1925). It is true that in 1830 no experiment showed the role of Planck's constant in mechanics.

5.3 Semi-Classical Approximation in Quantum Mechanics

The same idea can be applied to mechanics and the Schrödinger equation. This is called the semi-classical approximation of Brillouin, Kramers and Wentzel (BKW).

For example, we will refer to the book of Messiah [15] *Quantum mechanics*, volume 1, Chap. 6 for all details, in particular in the practical application of this method.

Consider the Schrdinger equation

$$i\hbar \frac{\partial}{\partial t} \psi(\mathbf{r}, t) = -\frac{\hbar^2}{2m} \Delta\psi(\mathbf{r}, t) + V(\mathbf{r})\,\psi(\mathbf{r}, t). \tag{5.40}$$

We separate in the wave function the modulus and the phase as

$$\psi(\mathbf{r}, t) = A(\mathbf{r}, t)\exp\left(\frac{i}{\hbar}S(\mathbf{r}, t)\right). \tag{5.41}$$

Substituting in (5.40) and separating the real part and imaginary part, we get

$$\frac{\partial S}{\partial t} + \frac{1}{2m}(\nabla S)^2 + V = \frac{\hbar^2}{2m}\frac{\nabla^2 A}{A} \tag{5.42}$$

$$m\frac{\partial A}{\partial t} + \nabla A \cdot \nabla S + \frac{1}{2}A\nabla^2 S = 0. \tag{5.43}$$

The second equation expresses the conservation of probability. If we introduce the probability density ρ and the current probability density $\mathbf{J}$ as

$$\rho(\mathbf{r}, t) = \psi^*(\mathbf{r}, t)\psi(\mathbf{r}, t) = A^2\ , \ \ \mathbf{J}(\mathbf{r}, t) = \frac{\hbar}{2im}(\psi^*\nabla\psi - \psi\nabla\psi^*) = \frac{A^2}{m}\nabla S\ ,$$

the conservation of probability is written in local form

$$\frac{\partial}{\partial t}\rho(\mathbf{r}, t) + \nabla \cdot \mathbf{J}(\mathbf{r}, t) = 0. \tag{5.44}$$

This equation amounts, with the form (5.41) and by multiplying (5.43) by $2A$, to

$$m\frac{\partial}{\partial t}A^2 + \nabla \cdot (A^2\nabla S) = 0. \tag{5.45}$$

This equation is closer to the Eq. (5.6).

The classical approximation is to take the limit $\hbar \to 0$ in Eq. (5.42) or

$$\frac{\partial S}{\partial t} + \frac{1}{2m}(\nabla S)^2 + V = 0, \tag{5.46}$$

which is the classical Hamilton-Jacobi equation.

Therefore, in the semi-classical approximation, the wave function can be considered as describing a conventional particle fluid without mutual interactions, in the potential V. The density and the current density of these particles are at all times equal to the quantum probability density ρ and current probability density $\mathbf{J}$.

5.4   Hamilton-Jacobi Formalism

Coming back to the Hamilton-Jacobi equation in a more mechanical mind.

$$\frac{\partial S}{\partial t} + H(x_1, \ldots, x_N, \frac{\partial S}{\partial x_1}, \ldots, \frac{\partial S}{\partial x_N}; t) = 0. \tag{5.47}$$

The Hamilton-Jacobi equation is a nonlinear partial differential equation, of the first order. In the same way as the equations of Lagrange-Euler or the canonical equations, it allows to calculate the motion. The use of these formalisms is a matter of convenience or mathematical structure of the problem.

The Hamilton-Jacobi equation is particularly suitable for the separation of variables and the choice of variables appropriate to the symmetry of a problem.

Jacobi's Theorem

An important result is useful: Jacobi's theorem, that we will explain on the simplest case of one dimension q.

Theorem *Let be a an integration constant, and suppose that we know the $S(q, a, t)$ action. So $\beta = \partial S/\partial a$ is a constant of the motion.*

Proof We have by definition

$$\beta = \frac{\partial S}{\partial a} \quad \text{that is} \quad \frac{d}{dt}\beta = \frac{\partial}{\partial t}\frac{\partial S}{\partial a} + \dot{q}\frac{\partial^2 S}{\partial q\,\partial a}. \tag{5.48}$$

Now, $\dot{q}$ is, by definition, the derivative of q *along the physical trajectory*, therefore

$$\dot{q} = \frac{\partial H}{\partial p}, \quad \text{and} \quad \frac{d}{dt}\beta = \frac{\partial}{\partial t}\frac{\partial S}{\partial a} + \frac{\partial H}{\partial p}\frac{\partial^2 S}{\partial q\,\partial a}. \tag{5.49}$$

Moreover, we have

$$p = \frac{\partial S(q, a, t)}{\partial q} \quad \text{therefore} \quad \frac{\partial}{\partial a}H(q, \frac{\partial S(q, a, t)}{\partial q}) = \frac{\partial H}{\partial p}\frac{\partial^2 S}{\partial a\partial q}. \tag{5.50}$$

By inserting in (5.48), we obtain the desired result

$$\frac{d}{dt}\beta = \frac{\partial}{\partial a}\left(\frac{\partial S}{\partial t} + H(q, \frac{\partial S(q, a, t)}{\partial q})\right) = 0, \tag{5.51}$$

on the trajectory, owing to the Hamilton-Jacobi equation (5.47).

Example: Central Potential

There are many applications of this equation, especially when variables can be separated.

Let us confine ourselves here, as an example, to a problem which encompasses the Kepler problem, and spherical coordinates.

In spherical coordinates (r, θ, ϕ) the Hamiltonian is written

$$H = \frac{1}{2m}\left(p_r^2 + \frac{p_\theta^2}{r^2} + \frac{p_\phi^2}{r^2 \sin^2 \theta}\right) + V(r, \theta, \phi). \tag{5.52}$$

Variables can be separated if the potential is of the form

$$V = V_0(r) + \frac{f(\theta)}{r^2} \tag{5.53}$$

(a term of the form $g(\phi)/r^2 \sin^2 \theta$ could also be added).

We set

$$S(\mathbf{r}, t) = S_0(\mathbf{r}) - Et \quad \text{that is} \quad \frac{\partial S}{\partial t} = -E, \tag{5.54}$$

where E is the constant value of energy. The Hamilton-Jacobi equation then becomes

$$\frac{1}{2m}\left(\frac{\partial S_0}{\partial r}\right)^2 + V_0(r) + \frac{1}{2mr^2}\left[\left(\frac{\partial S_0}{\partial \theta}\right)^2 + 2mf(\theta)\right] + \frac{1}{2mr^2 \sin^2 \theta}\left(\frac{\partial S_0}{\partial \phi}\right)^2 = E \tag{5.55}$$

The variable ϕ is cyclic, we note $\ell = L_z$ the constant value p_ϕ. In other words,

$$\left(\frac{\partial S_0}{\partial \phi}\right)^2 = \ell^2. \tag{5.56}$$

Carrying this into (5.55) reduces the problem to

$$\frac{1}{2m}\left(\frac{\partial S_0}{\partial r}\right)^2 + V_0(r) + \frac{1}{2mr^2}\left[\left(\frac{\partial S_0}{\partial \theta}\right)^2 + 2mf(\theta)\right] + \frac{\ell^2}{2mr^2 \sin^2 \theta} = E \tag{5.57}$$

Multiplying by $2mr^2$ we see that this equation separates in the sum of two terms, one related to the variable θ, the other to the variable r.

We are therefore looking for a solution of the form

$$S_0 = \ell\,\phi + S_1(\theta) + S_2(r). \tag{5.58}$$

We get

$$\left(\frac{dS_1}{d\theta}\right)^2 + 2mf(\theta) + \frac{\ell^2}{\sin^2\theta} = a, \tag{5.59}$$

$$\frac{1}{2m}\left(\frac{dS_2}{dr}\right)^2 + V_0(r) + \frac{a}{2mr^2} = E, \tag{5.60}$$

where a is, like E and ℓ, a constant of the motion, determined by the initial conditions.

The integration of these equations gives

$$S = -Et + \ell\,\phi + \int\sqrt{\left(2m(E - V_0(r)) - \frac{a}{r^2}\right)}\,dr + \int\sqrt{\left(a - 2mf(\theta) - \frac{\ell^2}{\sin^2\theta}\right)}\,d\theta \tag{5.61}$$

where (E, ℓ, a) are arbitrary integration constants .

To obtain the equations of motion, we use Jacobis theorem.

Let's resume the result (5.61) and consider the three (E, ℓ, a) constants of the motion. From the expression (5.61) of the action, the three constants β_E, β_ℓ, β_a are defined by

$$\beta_E = \frac{\partial S}{\partial E}, \quad \beta_\ell = \frac{\partial S}{\partial \ell}, \quad \beta_a = \frac{\partial S}{\partial a}.$$

The value of these constants is fixed by the initial conditions of the problem. We thus obtain the trajectory and the equation of the motion (5.61) from the three equations in one variable obtained from E, ℓ and a.

5.5 Exercises

5.1. The Lorentz Hamiltonian

Prove that, with the Hamiltonian (4.39), Hamilton's equations give the expected equation of motion.

5.2. Virial Theorem

We consider, in three dimensions, a particle of mass m placed in a potential $V(\mathbf{r})$, whose Hamiltonian is $H = p^2/2m + V(\mathbf{r})$. We assume that the particle is in a bound state with a given energy E.

1. Consider the physical quantity $A = \mathbf{r} \cdot \mathbf{p} \equiv xp_x + yp_y + zp_z$. Calculate the Poisson bracket $\{A, H\}$. Deduce from that the time evolution of A in terms of the variables $\mathbf{r}$ and $\mathbf{p}$.
2. We assume that the motion of the particle is periodic of period T. Let $f(\mathbf{r}, \mathbf{p})$ be a physical quantity, we define its mean value $\langle f \rangle$ by

$$\langle f \rangle = \frac{1}{T} \int_0^T f(t)\, dt \tag{5.62}$$

Considering the mean value of $\dot{A} \equiv dA/dt$, show that we have

$$2\langle \frac{p^2}{2m} \rangle = \langle \mathbf{r} \cdot \nabla V \rangle \tag{5.63}$$

3. What does this equality become if the potential V is a central power law function $V = g\, r^n$ with $r = |\mathbf{r}|$?
4. In the above case, what is the relation between the total energy E, the mean kinetic energy $\langle E_k \rangle$ and the mean potential energy $\langle V \rangle$ for

   (a) a harmonic oscillator $n = 2$, and) for a Newtonian (or Coulomb) potential $n = -1$?
5. In general, for an arbitrary potential, the orbits of two-body bound states are not closed curves, but they nevertheless remain confined in space. At all times, $|\mathbf{r}| \leq r_0$ and $|\mathbf{p}| \leq p_0$ where r_0 and p_0 are fixed. Give a generalization of the definition (5.62) such that the result (5.63) remains true.

Chapter 6
Lagrangian Field Theory

The Lagrangian formalism acquires its real power when one deals with systems that possess a large, possibly infinite, number of degrees of freedom. That is the case in mechanics of continuous media. We will now examine how this formalism deals with field theory.

In itself, field theory is a vast domain that acquires its completeness when one considers the quantization of fields and the theory of fundamental interactions. We cannot ignore the practical importance of this subject in many present technologies, which range from the acoustics of concert halls to the many modes of communication, whether terrestrial, submarine, space, with vibratory modes, advanced optics and electronics. In this rather short chapter, we will only give the principles of Lagrangian field theory and its application to the electromagnetic field. In the present chapter we want to explain the principles of Lagrangian field theory and its application to the electromagnetic field. The classical theory of gravitation is beyond the scope of this book. It is thoroughly treated in the literature, and we refer the interested reader to references [2, 17, 18]), for instance.

In Sect. 5.1, we study the principle of the Lagrangian formulation of field theory, starting with the case of a vibrating string. Actually, the procedure is rather simple. One starts by considering a discrete problem with finite elements of the string. One then takes the continuum limit such that a *Lagrangian space density* appears. It is in this limiting procedure that one appreciates how well the Lagrangian formalism is adapted to this type of problem.

The extension to three space dimensions, as well as several degrees of freedom, is dealt with in Sect. 5.2. One can easily guess the extension of the method to four dimensional space-time and relativistic fields. In Sect. 5.3, we will consider a scalar field, and in Sect. 5.4 the electromagnetic field and the Maxwell equations. In Sect. 5.5, we shall say a few words about field equations that are of first order in time. The first example is the Fourier diffusion equation, which corresponds to a nonreversible problem; i.e., a dissipative problem. This example is interesting because of the sim-

© The Author(s), under exclusive license to Springer Nature Switzerland AG 2023 105
J.-L. Basdevant, *Variational Principles in Physics*,
https://doi.org/10.1007/978-3-031-21692-3_6

ilarity between the Fourier equation and the Schrödinger equation. We shall see that a Lagrangian approach can be constructed for the latter but that essentially it leads nowhere in nonrelativistic quantum mechanics.

6.1   Vibrating String

The vibrating string is the prototype of a system with an infinite number of degrees of freedom.

Consider an elastic string of length l, fixed horizontally between the endpoints $x = 0$ and $x = l$ (we do not take gravity into account). Its linear mass density ρ is assumed to be uniform.

We only consider deformations of the string in the transverse plane (transverse waves). We denote by $\psi(x, t)$ the transverse (vertical) displacement at point x with respect to its position at rest. For simplicity, we assume that this displacement occurs in a single direction (the vertical axis).

One can consider the string to be the set of a large number of elements of length dx, each of which obeys the usual laws of dynamics. In the limiting procedure, this will result in an infinite number of degrees of freedom.

Consider an element of length dx. Its kinetic energy is

$$dE_k = \frac{1}{2}(\rho \, dx)\left(\frac{\partial \psi}{\partial t}\right)^2 . \tag{6.1}$$

Let τ be the elasticity constant of the string. If the displacement of two successive elements located at x and $x + dx$ varies compared with its value at rest, the corresponding potential energy V varies by

$$dV = \tau\left(\sqrt{1 + \left(\frac{\partial \psi}{\partial x}\right)^2} - 1\right) dx,$$

where obviously $(\partial \psi / \partial x)^2 \ll 1$. The variation V of the potential energy of the string when it is deformed is therefore

$$V = \frac{1}{2}\tau \int_0^l \left(\frac{\partial \psi}{\partial x}\right)^2 dx. \tag{6.2}$$

The Lagrangian of the string is the sum of the elementary Lagrangians:

$$\mathcal{L} = \frac{1}{2}\int_0^l \left[\rho\left(\frac{\partial \psi}{\partial t}\right)^2 - \tau\left(\frac{\partial \psi}{\partial x}\right)^2\right] dx. \tag{6.3}$$

If we consider the string as an assembly of material elements of length dx, each of these has an elementary Lagrangian

$$d\mathcal{L} = L\left(\psi, \frac{\partial\psi}{\partial t}, \frac{\partial\psi}{\partial x}\right) dx = \frac{1}{2}\left[\rho\left(\frac{\partial\psi}{\partial t}\right)^2 - \tau\left(\frac{\partial\psi}{\partial x}\right)^2\right] dx. \qquad (6.4)$$

The quantity L that appears in this expression is called the *Lagrangian density* of the string. In fact, the action of the string is

$$S = \int L\, dx\, dt = \frac{1}{2}\int\left[\rho\left(\frac{\partial\psi}{\partial t}\right)^2 - \tau\left(\frac{\partial\psi}{\partial x}\right)^2\right] dx\, dt. \qquad (6.5)$$

(The integral over x runs on the path $[0, l]$.)

We see that this is now a two-dimensional problem (x, t) for the dynamical quantity $\psi(x, t)$. We must minimize the integral (6.5). The corresponding Lagrange–Euler equation is

$$\frac{\partial L}{\partial\psi} = \frac{\partial}{\partial t}\left(\frac{\partial L}{\partial(\partial\psi/\partial t)}\right) + \frac{\partial}{\partial x}\left(\frac{\partial L}{\partial(\partial\psi/\partial x)}\right). \qquad (6.6)$$

In the case under consideration, $\partial L/\partial\psi = 0$ so that if we define the propagation velocity c by

$$c^2 = \frac{\tau}{\rho}, \qquad (6.7)$$

we obtain the propagation equation of vibrations along the string

$$\left(\frac{\partial^2\psi}{\partial t^2}\right) - c^2\left(\frac{\partial^2\psi}{\partial x^2}\right) = 0. \qquad (6.8)$$

We therefore see how the problem of a wave propagation can be deduced from a variational principle. Here, the difference between the total kinetic energy of the string and its potential energy must be as small as possible.

6.2 Field Equations

6.2.1 Generalized Lagrange–Euler Equations

The previous case is slightly more complex than the equations we saw in (2.8) and (2.9). Indeed, for a field, the dynamical variable ψ depends on several variables. In the example (6.8), the field ψ depends on two variables, t and x.

More generally, consider n dynamical variables ψ_k, $k = 1, \ldots, n$, that depend on m variables x_s, $s = 1, \ldots, m$ (including time); i.e., $\psi_k(x_s)$, $s = 1, \ldots, m$.

We define

$$\psi_k^s \equiv \frac{\partial \psi_k}{\partial x_s},$$ (6.9)

and we denote by $[\psi_k^s]$ the set of partial derivatives of $\psi_k(x_1, \ldots, x_m)$. The Lagrangian density is of the form

$$L(\psi_1, [\psi_1^s], \ldots, \psi_n, [\psi_n^s])$$

and the action is

$$S = \int L(\psi_1, [\psi_1^s], \ldots, \psi_m, [\psi_m^s]) dx_1 \ldots dx_m.$$

It is a bit tedious but not difficult to convince oneself that the determination of the extremum of the action S under the set of all infinitesimal transformations $\psi_k \to \psi_k + \delta\psi_k$, $k = 1, \ldots, n$, which vanish on the edge of the integration volume once one has performed all integrations by parts, lead to the generalized Lagrange–Euler equations

$$\frac{\partial L}{\partial \psi_k} = \sum_{s=1}^{m} \frac{\partial}{\partial x_s} \left(\frac{\partial L}{\partial \psi_k^s} \right).$$ (6.10)

In relativistic field theory, it is natural to incorporate time t among the variables (x, y, z, t) on which the fields ψ_k depend. In many problems, for instance in statistical field theory, it is useful to maintain the special role of the time variable. If we define

$$\dot{\psi}_k \equiv \frac{\partial \psi_k}{\partial t},$$

we obtain

$$\frac{\partial}{\partial t} \left(\frac{\partial L}{\partial \dot{\psi}_k} \right) = \frac{\partial L}{\partial \psi_k} - \sum_{s=1}^{m-1} \frac{\partial}{\partial x_s} \left(\frac{\partial L}{\partial \psi_k^s} \right),$$ (6.11)

of which (6.6) is a particular case.

6.2.2 Hamiltonian Formalism

Consider again the vibrating string, adding for more generality a linear term in ψ (which can come from an external force $F(x)$ that we apply at each point). For simplicity, we define

$$\dot{\psi} \equiv \frac{\partial \psi}{\partial t} \quad \text{and} \quad \psi' \equiv \frac{\partial \psi}{\partial x},$$ (6.12)

and we consider the Lagrangian density

$$L = \frac{1}{2}[\rho(\dot{\psi})^2 - \tau(\psi')^2] + F\psi, \tag{6.13}$$

which leads to the equation of motion

$$\left(\frac{\partial^2 \psi}{\partial t^2}\right) - c^2 \left(\frac{\partial^2 \psi}{\partial x^2}\right) = G, \tag{6.14}$$

where $G = F/\rho$.

Since we are interested in the *time* evolution of the system, we define the density of conjugate momentum p by

$$p = \frac{\partial L}{\partial \dot{\psi}}, \quad \text{i.e., here} \quad p = \rho \frac{\partial \psi}{\partial t}. \tag{6.15}$$

For a vibrating string, this is the linear density of momentum.

The Hamiltonian density is

$$H = p\dot{\psi} - L = \frac{1}{2}\rho[(\dot{\psi})^2 + c^2(\psi')^2] - F\psi = \frac{p^2}{2\rho} + \frac{1}{2}\tau(\psi')^2 - F\psi. \tag{6.16}$$

This density depends on ψ and p, but also on ψ', and the form of the canonical equations must be modified. Inserting (6.16) (i.e., $L = p\dot{\psi} - H$), in the least action principle, and integrating by parts in the two variables x and t, we obtain

$$0 = \delta \int dt \int dx(p\dot{\psi} - H(p, \psi, \psi'))$$

$$= \int dt \int dx[\dot{\psi}\delta p + p\delta\dot{\psi} - (\partial H/\partial p)\delta p - (\partial H/\partial \psi)\delta\psi - (\partial H/\partial\psi')\delta\psi']$$

$$= \int dt \int dx \left[\left(\dot{\psi} - \frac{\partial H}{\partial p}\right)\delta p - \left(\dot{p} + \frac{\partial H}{\partial\psi} - \frac{\partial}{\partial x}\frac{\partial H}{\partial\psi'}\right)\delta\psi\right]. \tag{6.17}$$

Therefore, Hamilton's equations are

$$\frac{\partial \psi}{\partial t} = \frac{\partial H}{\partial p}; \quad \frac{\partial p}{\partial t} = \frac{\partial}{\partial x}\left(\frac{\partial H}{\partial\psi'}\right) - \frac{\partial H}{\partial\psi}. \tag{6.18}$$

One can check that they yield the propagation equation (6.14).

6.3 Scalar Field

The previous results allow us to understand the form of the Lagrangian of a scalar field in three-dimensional space, for instance sound waves in a compressible nonviscous fluid. Calling $\psi(\mathbf{r}, t)$ the compression of the fluid, and c the sound velocity in the fluid, the Lagrangian density has the form

$$L = \frac{1}{2}\rho \left[(\nabla \psi)^2 - \frac{1}{c^2} \left(\frac{\partial \psi}{\partial t} \right)^2 \right]. \tag{6.19}$$

Notice that, compared to the vibrating string, space and time derivatives are interchanged. The kinetic term (local velocity) comes from a vector quantity, whereas the potential (the pressure) is a scalar.

With the Lagrangian density (6.19), one obtains the propagation equation

$$\frac{1}{c^2} \frac{\partial^2 \psi}{\partial t^2} - \Delta \psi = 0. \tag{6.20}$$

6.4 Electromagnetic Field

The case of the electromagnetic field is more complex and deeper. In fact, it involves two vector fields, and above all, we must take care of relativistic invariance, which is the fundamental property of Maxwell's equations. This problem is treated thoroughly in the book of Landau and Lifshitz [2], for instance. Here we want to point out the major features.

Physically, the electromagnetic field cannot be separated from its sources, the charges, on which it acts. For a system of charged particles in an electromagnetic field, the action is written in full generality as

$$S = S_{field} + S_{part} + S_{int}, \tag{6.21}$$

where S_{field} is the action of free fields, S_{part} is the action of the free particles in the absence of fields, and S_{int} corresponds to the interaction of these particles and the field, which we know already from Sect. 3.3.2. An electromagnetic field derives from the potentials $\mathbf{A}$ and Φ, the Lagrangian of a particle of charge q and mass m is expressed in terms of the *potentials* $\mathbf{A}$ and Φ,

$$\mathcal{L}_{int} = q\,\dot{\mathbf{r}} \cdot \mathbf{A}(\mathbf{r}, t) - q\,\Phi(\mathbf{r}, t). \tag{6.22}$$

Relativistic Field

The previous form (6.22) has the right Lorentz transformation properties.

We said in Chaps. 2 and 4 that Minkowski's space-time is based on a four-dimensional vector space with a Lorentz scalar product of signature $(+, -, -, -)$. An orthogonal base, in this convention, consists of four-vectors: $\mathbf{e}_i, i = 0, 1, 2, 3$ that have orthogonal relationships

$$\mathbf{e}_i.\mathbf{e}_j = g_{ij} \quad \text{with} \quad g_{00} = +1, \quad g_{ij} = -\delta_{ij}, \quad g_{ij} = 0; \quad \text{if; } i \neq j.$$

($g_{\mu\nu}$ is the metric tensor.)

We note $x^i = (x^0, x^1, x^2, x^3) = (ct, x, y, z) \equiv (ct, \mathbf{r})$ the (t, x, y, z) components of a *four-vector* of space-time, the spatial part is a usual vector of $\mathcal{R}^3$. A four-vector is a vector of Minkowski space, on which the changes of reference system are made by Lorentz transformations. The scalar product of two four-vectors a^i and b^i is, according to the Einstein convention of summation on the repeated up and down indices, a relativistic invariant

$$g_{ij}a^i b^j = a^0 b^0 - \mathbf{a}.\mathbf{b}. \tag{6.23}$$

In electromagnetism, we introduce the current four-vector

$$\{j^\mu\} = (c\rho, \mathbf{j}), \tag{6.24}$$

where ρ and $\mathbf{j}$ are respectively the charge density and current density of particles, and the potential four-vector field is

$$\{A^\mu\} = (\Phi/c, \mathbf{A}), \tag{6.25}$$

the *Lagrangian density* that corresponds to (6.22) is

$$\mathcal{L}_{int} = -g_{\mu\nu}j^\nu A^\mu \equiv -j_\mu A^\mu, \tag{6.26}$$

which is obviously invariant. We have seen that it is easier to work with potentials than with the fields themselves, whose transverse parts are mixed in Lorentz transformations.

(We keep the same symbol $\mathcal{L}$ for the Lagrangian density; the integration runs along space and time.) The action is invariant since $d^3r \, dt$ is a relativistic invariant.

Electromagnetic Tensor Field

The fields are expressed in terms of the potentials Φ and $\mathbf{A}$ by

$$\mathbf{B} = \nabla \times \mathbf{A}, \qquad \mathbf{E} = -\nabla \Phi - \frac{\partial \mathbf{A}}{\partial t}. \tag{6.27}$$

Using the notation $\partial^\mu = \partial/\partial x_\mu$, one expresses the electromagnetic tensor field as

$$F^{\mu\nu} = \partial^\mu A^\nu - \partial^\nu A^\mu; \tag{6.28}$$

i.e., the antisymmetric tensor

$$F^{\mu\nu} = \begin{pmatrix} 0 & -E_x/c & -E_y/c & -E_z/c \\ E_x/c & 0 & -B_z & B_y \\ E_y/c & B_z & 0 & -B_x \\ E_z/c & -B_y & B_x & 0 \end{pmatrix}.$$

The couple of homogeneous Maxwell equations follows from the structure of the tensor $F^{\mu\nu}$, and the four equations (or identities)

$$\partial^\mu F^{\nu\rho} + \partial^\nu F^{\rho\mu} + \partial^\rho F^{\mu\nu} = 0, \tag{6.29}$$

which lead to

$$\nabla \times \mathbf{E} = -\frac{\partial \mathbf{B}}{\partial t}, \qquad \nabla \cdot \mathbf{B} = 0. \tag{6.30}$$

The relationship between the sources and the field (second pair of Maxwell equations)

$$\nabla \cdot \mathbf{E} = \frac{\rho}{\epsilon_0} \qquad c^2 \nabla \times \mathbf{B} = \frac{\mathbf{j}}{\epsilon_0} + \frac{\partial \mathbf{E}}{\partial t}, \tag{6.31}$$

is

$$\partial^\mu F^{\mu\nu} = \frac{j^\nu}{\epsilon_0 c^2} = \mu_0 j^\nu.$$

From this tensor and its covariant conjugate $F_{\mu\nu}$, we can construct two *relativistic invariants*:

$$F_{\mu\nu} F^{\mu\nu} = -\frac{2}{c^2}(\mathbf{E}^2 - c^2 \mathbf{B}^2) \quad \text{and} \quad \epsilon_{\mu\nu\rho\sigma} F^{\mu\nu} F^{\rho\sigma} = -\frac{8}{c} \mathbf{E} \cdot \mathbf{B}, \tag{6.32}$$

where $\epsilon_{\mu\nu\rho\sigma}$, the Levi-Civita tensor, is equal to $+1$ for a even permutation of $\mu\nu\rho\sigma$, to -1 for an odd permutation and to 0 otherwise. The second invariant is the transverse orientation of fields and does not contribute to the energy.

Field and Matter

The inhomogeneous Maxwell equations relate the fields to the charge and current densities.

Suppose there is a given charge and current density $\{j^\mu\} = (c\rho, \mathbf{j})$, the density of electromagnetic field lagrangian and interaction lagrangian in the presence of these sources is therefore

$$\mathcal{L} = -j_\mu A^\mu - \frac{\varepsilon_0 c^2}{4} F_{\mu\nu} F^{\mu\nu} \quad \text{with} \quad \varepsilon_0 \mu_0 c^2 = 1. \tag{6.33}$$

The action S is defined as the integral over time and all space

$$S = \int \mathcal{L} \, d^3\mathbf{r} \, dt. \tag{6.34}$$

which is Lorentz invariant because the $d^3\mathbf{r} \, dt$ four-volume element is invariant.

Returning to an expression not manifestly covariant,

$$\mathcal{L} = -\rho\phi + \mathbf{j} \cdot \mathbf{A} + \frac{\varepsilon_0}{2}(E^2 - c^2 B^2), \tag{6.35}$$

We have

$$\frac{\partial \mathcal{L}}{\partial(\partial^\mu A^\nu)} = -\varepsilon_0 c^2 F_{\mu\nu} = \varepsilon_0 c^2 F_{\nu\mu}. \tag{6.36}$$

We can see that the equations of motion of the electromagnetic field are, in covariant form,

$$\partial_\mu F^{\mu\nu} = \mu_0 j^\nu. \tag{6.37}$$

These equations are invariant in a *gauge transformation*:

$$A_\mu \longrightarrow A_\mu + \partial_\mu \chi \tag{6.38}$$

which leaves invariant the $F_{\mu\nu}$ components of the electromagnetic tensor. This allows to impose an additional condition on the A_μ, for instance the *Lorentz gauge*

$$\partial_\mu A^\mu = 0. \tag{6.39}$$

In this choice of gauge, Maxwell's equations boil down to

$$\partial^\mu \partial_\mu A^\nu \equiv (\frac{1}{c^2} \frac{\partial^2}{\partial t^2} - \nabla^2) A^\nu = \mu_0 j^\nu. \tag{6.40}$$

In the non-relativistic formalism, this amounts as expected to

$$\nabla \cdot \mathbf{E} = \frac{\rho}{\varepsilon_0}, \qquad c^2 \nabla \times \mathbf{B} = \frac{\mathbf{j}}{\varepsilon_0} + \frac{\partial \mathbf{E}}{\partial t}. \tag{6.41}$$

We see from (6.35) that the physical electromagnetic field in the vacuum, away from charges, minimizes the difference $(E^2 - c^2 B^2)$ given the constraints imposed by the presence of the sources. This was implicit in the example of the simple electrostatic field in (2.3.1).

6.5 Equations of First Order in Time

In order to deal with equations of first order in time, such as the Fourier diffusion equation or the Schrödinger equation, we use the technique described in Sect. 3.3.1 for dissipative systems.

6.5.1 Diffusion Equation

Diffusion, be it of heat or of a substance in a medium, is nonreversible. In that sense, it can be thought of as a dissipative system. A quantity of heat placed at some point in a material diffuses in the material and tends to make its distribution as uniform as possible. It never "reconcentrates" at its initial position.

The technique developed in Sect. 3.3.1 allows us to formulate this in a Lagrangian form.

Let $\psi(\mathbf{r}, t)$ be the density of heat (or of a diffusing substance) and a^2 the diffusion constant. We introduce a fictitious mirror system whose density ψ^* "concentrates" instead of diffusing.

Consider the Lagrangian density

$$L = -\nabla \psi \cdot \nabla \psi^* - \frac{a^2}{2} \left(\psi^* \frac{\partial \psi}{\partial t} - \psi \frac{\partial \psi^*}{\partial t} \right); \tag{6.42}$$

i.e., the action is

$$S_{(t_0, t_1)} = \int_{t_0}^{t_1} dt \int L \, d^3\mathbf{r}.$$

The Lagrange–Euler equations give

$$\Delta \psi = a^2 \left(\frac{\partial \psi}{\partial t} \right) \quad \text{and} \quad \Delta \psi^* = -a^2 \left(\frac{\partial \psi^*}{\partial t} \right). \tag{6.43}$$

The equation satisfied by ψ is the usual diffusion equation. That written for ψ^* would represent a diffusion reversed in time, or a "concentration".

It is necessary to use similar techniques in order to write in a Lagrangian form the flow of a viscous fluid (see [12], Chap. 3, Sect. 3).

6.5.2 Schrödinger Equation

The Schrödinger equation is not a dissipative system since there is conservation of the norm and wave propagation. Nevertheless, the formal similarity between its structure and that of the Fourier equation[1] allows us to write a Lagrangian formulation similar to what we developed above.

We consider the simple case of a particle of mass m placed in a potential V. Here, the wave function ψ is complex. Therefore, one can simply use its complex conjugate ψ^* as the "mirror" dynamical variable. This amounts to considering the real and imaginary parts of the wave function as independent dynamical variables. In direct analogy with (6.42), the Lagrangian density is

$$L = -\frac{\hbar^2}{2m}\nabla\psi \cdot \nabla\psi^* - \frac{\hbar}{2i}\left(\psi^*\frac{\partial\psi}{\partial t} - \psi\frac{\partial\psi^*}{\partial t}\right) - \psi^*V\psi. \tag{6.44}$$

The Lagrange–Euler equations give

$$-\frac{\hbar^2}{2m}\Delta\psi + \frac{\hbar}{i}\frac{\partial\psi}{\partial t} = -V\psi \quad \text{and} \quad -\frac{\hbar^2}{2m}\Delta\psi^* - \frac{\hbar}{i}\frac{\partial\psi^*}{\partial t} = -V\psi^*, \tag{6.45}$$

as one can easily check.

The densities of conjugate momenta are

$$p = \frac{\partial L}{\partial\dot\psi} = -\frac{\hbar\psi^*}{2i}; \quad p^* = \frac{\partial L}{\partial\dot\psi^*} = \frac{\hbar\psi}{2i}, \tag{6.46}$$

and the Hamiltonian density is

$$H = \frac{\hbar^2}{2m}\nabla\psi \cdot \nabla\psi^* + \psi^*V\psi. \tag{6.47}$$

This form is appealing since its integral over space is simply the expectation value of the quantum energy

$$\int H \, d^3\mathbf{r} = \langle E \rangle = \int \left[-\frac{\hbar^2}{2m}\psi^*\Delta\psi + \psi^*V\psi\right] d^3\mathbf{r}. \tag{6.48}$$

---

[1] One says that the Schrödinger equation is a Fourier equation with an imaginary time.

It is therefore tempting to look for a variational principle analogous to what we saw in Sect. 2.3.1 for the electrostatic potential.

Unfortunately, this cannot be done. The Hamiltonian formulation of this problem is even more involved than in Sect. 6.2.2, and it does not bring anything new, compared with the Lagrangian formulation (6.44).

In fact, the problem lies in the form of the conjugate momenta (6.46). As one notices immediately, p and p^* are not independent variables. These momenta are proportional to the dynamical variables ψ and ψ^*. This property, however, is useful in quantum field theory, in particular in what one calls second quantization. This falls outside the scope of this book.

We will see the variational formulation of quantum mechanics via *path integrals* in Chap. 9.

6.6   Problem

Neutron transport in matter

The equation for neutron transport in matter has the form

$$a^2 \frac{\partial \rho}{\partial t} + \frac{3}{v^2} \frac{\partial^2 \rho}{\partial t^2} - \Delta \rho = 0, \tag{6.49}$$

called the *telegraph equation* (see, for instance, Appendix D of [16]) which shows a propagation term of the neutron density, of individual velocities v which we assume to be the same and constant here. In the diffusive regime, in reactor cores, this term is negligible. There exist situations (for instance, neutrino transport in supernovae) where all terms must be kept owing to the discontinuities of the diffusive medium.

Proceeding as in (3.40), write the form of a Lagrangian from which this equation is derived.

Chapter 7
Motion in a Curved Space

You cannot teach a crab to walk straight.
Aristophanes

7.1 The Equivalence Principle

Albert Einstein's (1879–1955) masterpiece, *general relativity* is based on the observation that two physical quantities that have a priori no relationship, are equal (or strictly proportional). This is, as we know, the two meanings of the concept of mass. One is that of *coefficient of inertia or acceleration resistance* of a body in the laws of dynamics, the other is that of *coupling coefficient to the gravitational field*. There is no a priori reason for this equality called the *Equivalence principle*. The motion of a charged particle in an electromagnetic field depends on the two independent parameters of mass, coefficient of inertia, and particle charge, coefficient of field coupling. In a gravitational field, the equality of the inertial mass and from the gravitational mass removes the mass of a body of the equations of motion. Two bodies placed in the same initial conditions have the same motion, regardless of their masses.

The depth of this observation was realized quite late by the scientific community. The historical experience of Eötvös (1848–1919) in 1890[1] has been repeated systematically. It is still performed with increasingly sophisticated techniques.

The underlying idea of General relativity is that the equality becomes natural if what we call the "gravitational" motion is actually a *free* motion in a *curved* space-time.

---

[1] Roland v. Eötvös, Mathematische und Naturwissenschaftliche Berichte aus Ungarn, **8**, 65, 1890.

© The Author(s), under exclusive license to Springer Nature Switzerland AG 2023
J.-L. Basdevant, *Variational Principles in Physics*,
https://doi.org/10.1007/978-3-031-21692-3_7

117

Einstein used to say[2] that in 1907, when he was working on how to incorporate Newtonian gravitation in relativity (the incorporation of electromagnetism was by construction automatic), he had the "happiest thought of his life" (the original version is "glücklichster Gedanke meines Lebens"). He was thinking of what someone falling from the roof would feel. For such an "observer" (and of course as long as they does not encounter any obstacle) *there is no gravitational field* (the italics are from Einstein). If such observers let any object "fall" from their pocket, this object stands still or has a uniform linear motion with respect to them, whatever its nature, and its physical and chemical composition; (the resistance of the atmosphere is neglected).

The basic idea of the equivalence principle and its consequences is explained in many texts including those of David Langlois [17], of Nathalie Deruelle and Jean-Philippe Uzan [18], and Hans Stefani [20]. The ambition of this chapter is to show how the notion of motion in a curved space can lead to a theory such that the equality of the "two masses" emerges naturally.

The equivalence principle can be stated in the following way.

For a short time, the laws of physics in a small laboratory in free fall are the same as they would be in the same laboratory in an inertial reference frame in the absence of gravitation.

One usually makes a distinction between this principle, which only concerns the motion, and the theory of general relativity itself, i.e. the Einstein equations that relate the curvature tensor of space-time to the energy momentum tensor of matter. In this book, we shall not describe fully Einstein's equations and their consequence.

Here we will discuss various initial applications of the principle of equivalence, such as precession of Mercury's perihelion and the deviation of light rays by the gravitational field. In the next chapter we will describe a great discovery of fundamental research: the detection of gravitational waves in September 2015. This natural phenomenon was established one century after it was predicted by Albert Einstein.

We will start by studying the free motion of a particle in a curved space. In Sect. 7.1, we define what one calls a curved space and introduce the fundamental notion of the *metric* of the space. In Sect. 7.2, we will write the motion of a free particle in such a space. This will lead us, in Sect. 7.3, to a fundamental result: The physical trajectories are the *geodesics* of the space; i.e., the curves of minimal (or extremal) length. As we shall see, this is how the motion of a particle of constant energy E in a Euclidean space-time, can be transformed into the *free* motion of a similar particle in a curved space, which is equivalent to the Maupertuis principle.

This will allow us to understand in Sect. 7.4 the reasoning of Einstein when he constructed general relativity, and some consequences of this theory. We will display, three historical examples: the variation of the beat of a clock due to the gravitational field, the corrections to Newton's celestial mechanics, and the deviation of light rays by a gravitational field.

---

[2] See, for instance, A. Pais, *Subtle Is the Lord*, Chapter 9, Oxford University Press, New York, 1982. The original letter of Einstein to R. W. Lawson in January 1920 has been found. The published article, A. Einstein, *Nature*, **106**, 782 (1921), is not as light in spirit.

These examples are historical. As we shall see in Sect. 7.5, they are also very important in present-day astrophysics and cosmology. The deviation of light by a gravitational field plays an important role via the gravitational lensing effect that it induces. One application is the search for a baryonic component in the "missing mass" of the universe. Another is that the mass distribution in the universe, be it the visible mass or the missing mass, acts as a natural telescope that can enable us to see faraway objects, and therefore much younger objects. Through this natural cosmic telescope (or microscope), the universe appears as an endless gallery of gravitational mirages.

7.2 Curved Spaces

7.2.1 Generalities

It is the work of mathematicians on Euclid's fifth axiom (the postulate of parallel lines) that led to the developments on the existence and properties of non-Euclidean spaces in the 19th century. Legendre (1752–1833) had shown that Euclid's axiom is equivalent to the assumption that the sum of the angles of a triangle is equal to π. As early as 1816, Carl Friedrich Gauss (1777–1855) had convinced himself that this statement could not be proven (roughly 15 years before the celebrated work of Nikolay Ivanovitch Lobatchevsky).

Gauss then addressed the question of *measuring* whether the three-dimensional space in which we live is "flat" (i.e. Euclidean) or not. Gauss tried to verify that the sum of the angles of a triangle is equal to 180°. He performed this measurement between three peaks of the Harz, in central Germany: the Inselberg, the Brocken and the Hoher Hagen. As "straight lines," he used light rays reflected between these three points, and he had to conclude that (unfortunately) the sum of the angles is equal to 180°. Some remarks are in order on the idea and on the result.

1. If the accuracy of the measurement had been good enough (owing to atmospheric density fluctuations, accurate measurements could not be performed with present laser technology), Gauss would have detected a slight deviation ($\sim 10^{-10}$), and therefore a curvature of space, since light rays are deviated by the Earth's gravitational field.

2. A measurement, the spirit of which is close to the idea of Gauss, was performed in 1964 by Shapiro,[3] who measured the delay of a radar echo between the Earth and space probes placed on the planets Mercury, Venus, and Mars (the *Viking* program). When the planet crosses the direction of the sun opposite to the earth, the delay is longer than what one would expect from Euclidean geometry. (Celestial mechanics allows us to calculate the orbits, and therefore the Euclidean distances, with great accuracy.)

---

[3] I.I. Shapiro, Phys. Rev. Lett., **13**, 789 (1964).

3. Gauss's idea was similar to that of Eratosthenes when he measured the radius of the Earth by comparing the shadows of two vertical sticks on the same meridian, in Syene and in Alexandria, on the summer solstice. Eratosthenes had read in a document the observation that on the day of the summer solstice, and only on that day, the wells in Syene (Aswan), on the tropic of Cancer, had no shadow inside at noon. At any other moment, a shadow appeared somewhere on the sides and on the bottom. Eratosthenes concluded that the sun was, at that moment, at the vertical of Syene. He measured the shadow of a vertical bar in Alexandria at noon on the same day on the same meridian.

The measurement gave him an angle of 7° and 12 min. He figured out the distance between Syene and Alexandria (probably the most difficult task in the experiment) and found a value of the circumference of the order of 40,350 km (compared with the actual value of 40,074 km). There is part luck in the accuracy of the value (Eratosthenes did not know how far each city was from the same meridian). However, intellectually, the experiment is fascinating since it provides a means to probe the structure of the space in which we live and to measure its radius.

7.2.2   The Light Rays, Geodesics of Our Space

4. The necessary tool for this type of measurement is to have *straight lines* (i.e., *geodesics*) of the space. It appears that always, whether it was Thales measuring the height of the Great Pyramid, Eratosthenes, or Gauss, it was implicit in the minds of people that light rays are *physical* entities that possess the "perfect" mathematical property of propagating along straight lines.

In his celebrated memoir on the theory of surfaces, Gauss understood that the geometry of a surface is an *intrinsic* property of the surface, independent of whether this surface is embedded in a Euclidean space or not. Gauss's ideas were the starting points of the developments performed by Riemann.

In order to see whether or not a space is Euclidean, one can check whether the Pythagorean theorem, the triangle inequality, and the angle formula above are satisfied or not. Analyzing this further shows that everything boils down to measuring *distances* and comparing sets of them. Hence the importance of what is called the *metric tensor* or simply the *metric* of the space, which we shall introduce below.

A famous example, due to Einstein, illustrates this fact. Consider four points in a space, which we denote 1, 2, 3, 4, and let us denote d_{ij} the distance between points i and j. In a flat, Euclidean, space the following relation is always satisfied

$$d_{12}^4 d_{34}^2 + d_{13}^4 d_{24}^2 + d_{14}^4 d_{23}^2 + d_{23}^4 d_{14}^2 + d_{24}^4 d_{13}^2 + d_{34}^4 d_{12}^2$$
$$+ d_{12}^2 d_{23}^2 d_{31}^2 + d_{12}^2 d_{24}^2 d_{41}^2 + d_{13}^2 d_{34}^2 d_{41}^2 + d_{23}^2 d_{34}^2 d_{42}^2$$
$$- d_{12}^2 d_{23}^2 d_{34}^2 - d_{13}^2 d_{32}^2 d_{24}^2 - d_{12}^2 d_{24}^2 d_{43}^2 - d_{14}^2 d_{42}^2 d_{23}^2$$
$$- d_{13}^2 d_{34}^2 d_{42}^2 - d_{14}^2 d_{43}^2 d_{32}^2 - d_{23}^2 d_{31}^2 d_{14}^2 - d_{21}^2 d_{13}^2 d_{34}^2$$
$$- d_{24}^2 d_{41}^2 d_{13}^2 - d_{21}^2 d_{14}^2 d_{43}^2 - d_{31}^2 d_{12}^2 d_{24}^2 - d_{32}^2 d_{21}^2 d_{14}^2 = 0.$$

One can use an airline schedule (and some courage) to verify that this equality is not satisfied by Paris, New York, Johannesburg, and Shanghai (or any other set of four airports), provided one uses the *actual distances* covered by airplanes going as "straight" as possible from one place to the other.

7.2.3 Metric Tensor

We characterize a point of the space[4] by a set of coordinates $\{x^\alpha\}$. The distance ds between two infinitesimally separated points $\{x^\alpha\}$ and $\{x^\alpha + dx^\alpha\}$ is given, by the definition of the metric tensor $g_{\alpha\beta}$ of the space, as

$$ds^2 = \sum_{\alpha,\beta} g_{\alpha\beta} dx^\alpha \, dx^\beta \equiv g_{\alpha\beta} dx^\alpha \, dx^\beta. \qquad (7.1)$$

In the second form, we make use of Einstein's convention of summation over repeated indices.

The inverse $g^{\alpha\beta}$ is defined by

$$g^{\alpha\nu} g_{\nu\beta} = \delta^\alpha_\beta. \qquad (7.2)$$

In Euclidean space $\mathcal{R}^3$, in Cartesian coordinates, we have

$$ds^2 = dx^2 + dy^2 + dz^2, \text{ i.e., } g_{xx} = g_{yy} = g_{zz} = 1, \text{ and } g_{ij} = 0 \text{ otherwise.}$$

In spherical coordinates (r, θ, ϕ), the metric tensor is also diagonal, but its elements are no longer constants:

$$g_{rr} = 1 \, , \ g_{\theta\theta} = r^2 \, , \ g_{\phi\phi} = r^2 \sin^2 \theta.$$

---

[4] Or the manifold; we use both terms.

7.2.4 Examples

Sphere S^2 in $\mathcal{R}^3$

Let (x, y, z) be the coordinates of a point of three-dimensional Euclidean space $\mathcal{R}^3$. A sphere of radius R centered at the origin corresponds to points such that $x^2 + y^2 + z^2 = R^2$. The square of the distance between two (infinitesimally distant) points in $\mathcal{R}^3$ is $ds^2 = dx^2 + dy^2 + dz^2$. On the sphere, we have of course $z\,dz = -(x\,dx + y\,dy)$ so that, putting it all together, we have in Cartesian coordinates

$$ds^2 = dx^2 + dy^2 + \frac{(x\,dx + y\,dy)^2}{R^2 - x^2 - y^2};$$

i.e., the metric tensor

$$g_{xx} = 1 + \frac{x^2}{R^2 - x^2 - y^2} \quad g_{yy} = 1 + \frac{y^2}{R^2 - x^2 - y^2}, \quad g_{xy} = g_{yx} = \frac{xy}{R^2 - x^2 - y^2}.$$

Of course, the expression is much simpler in spherical coordinates (θ, ϕ):

$$ds^2 = R^2(d\theta^2 + \sin^2\theta d\phi^2); \quad \text{i.e.,} \quad g_{\theta\theta} = R^2, \quad g_{\phi\phi} = R^2 \sin^2\theta.$$

Three-Dimensional Spaces Embedded in Four-Dimensional Euclidean Space $\mathcal{R}^4$

Similarly, we can construct isotropic three-dimensional curved spaces embedded in $\mathcal{R}^4$. In addition to the usual metric of a flat space, there is a curvature term, which is particularly simple to express in spherical coordinates (ρ, θ, ϕ). Let (x, y, z, w) be the Cartesian coordinates in $\mathcal{R}^4$. We then obtain the following:

1. "Spherical" space, S^3 sphere:

$$w^2 + x^2 + y^2 + z^2 = R^2; \quad \text{i.e.,} \quad dw^2 = \frac{\rho^2 d\rho^2}{R^2 - \rho^2}. \tag{7.3}$$

2. Hyperbolic spaces:
   Two-sheet hyperboloid:

$$w^2 - (x^2 + y^2 + z^2) = R^2; \quad \text{i.e.,} \quad dw^2 = \frac{\rho^2 d\rho^2}{R^2 + \rho^2}, \tag{7.4}$$

   One-sheet hyperboloid:

$$(x^2 + y^2 + z^2) - w^2 = R^2; \quad \text{i.e.,} \quad dw^2 = \frac{\rho^2 d\rho^2}{\rho^2 - R^2}. \tag{7.5}$$

3. Parabolic space:

$$w - \frac{(x^2 + y^2 + z^2)}{2a} = 0; \quad \text{i.e.,} \quad dw^2 = \frac{\rho^2 d\rho^2}{a^2}. \tag{7.6}$$

In all these cases, the metric tensor is expressed as

$$ds^2 = (1 + f(\rho)^2)d\rho^2 + \rho^2 d\theta^2 + \rho^2 \sin^2 \theta d\phi^2. \tag{7.7}$$

General Case

One could, of course, continue playing the same type of game as in these examples by imposing any constraint of the type $\Phi(x, y, z, w) = 0$ in the space $\mathcal{R}^4$. Actually, one would be far from discovering all three-dimensional curved spaces.

The definition of a curved space consists of *choosing* the metric $\{g_{\alpha\beta}\}$; the simple examples above are only illustrations.

Historically, the most famous example was given by Felix Klein in 1890. It was a concrete example of the geometries of Gauss, János Bólyai, and Lobatchevsky. Klein's "bottle" consists of an analytical geometry where each point is represented by two real numbers, x_1 and x_2, such that $x_1^2 + x_2^2 < 1$ and where the distance $d(x, y)$ between two points is defined to be

$$\cosh\left[\frac{d(x, y)}{a}\right] = \frac{1 - x_1 y_1 - x_2 y_2}{\sqrt{(1 - x_1^2 - x_2^2)(1 - y_1^2 - y_2^2)}}, \tag{7.8}$$

where a is a dimensional scale parameter.

Note that this two-dimensional space, whose curvature is negative (as opposed to a sphere, which has a positive curvature) cannot be embedded in three-dimensional Euclidean space $\mathcal{R}^3$. It can only be embedded in spaces of dimension greater than three.

7.3 Free Motion in a Curved Space

We now study the free motion of a particle of mass m in a curved space.

7.3.1 Lagrangian

Since the particle is free, the Lagrangian boils down to its kinetic part, $E_{kin} = mv^2/2$; i.e.,

$$\mathcal{L} = \frac{1}{2}m\left(\frac{ds}{dt}\right)^2 = \frac{1}{2}mg_{\alpha\beta}\frac{dx^\alpha}{dt}\frac{dx^\beta}{dt}. \tag{7.9}$$

Note that if the space variables do not seem to appear explicitly in this Lagrangian, they are present in the metric $g_{\alpha\beta}$.

The conjugate momenta are obtained with no difficulty. Assuming the metric is symmetric, $g_{\alpha\beta} = g_{\beta\alpha}$, which does not restrain the generality, one obtains

$$p_\alpha = \frac{\partial\mathcal{L}}{\partial\dot{x}^\alpha} = mg_{\alpha\beta}\frac{dx^\beta}{dt}. \tag{7.10}$$

The Hamiltonian is

$$H = p_\alpha\dot{x}^\alpha - \mathcal{L} = \mathcal{L}. \tag{7.11}$$

The value of the Hamiltonian is the same as the value of the Lagrangian as it should be since we consider a free particle. (Of course, the Lagrangian and Hamiltonian functions are not expressed with the same variables.) We deduce a consequence which is both obvious and important. Because of energy conservation, the square of the velocity is a *constant of the motion* along the trajectory.

$$\frac{d}{dt}\left(\frac{1}{2}mv^2\right) = \frac{d}{dt}\mathcal{L} = \frac{d}{dt}H = 0 \quad . \tag{7.12}$$

This property is the curved-space extension of Galileo's principle of inertia. In the particular case of a Euclidean space, the velocity vector of a particle is constant.

7.3.2 Equations of Motion

The equations of motion are obtained in the usual way

$$\frac{\partial\mathcal{L}}{\partial x^\nu} = \frac{1}{2}m\frac{\partial g_{\alpha\beta}}{\partial x^\nu}\dot{x}^\alpha\dot{x}^\beta = \frac{d}{dt}\left(\frac{\partial\mathcal{L}}{\partial\dot{x}^\nu}\right) = m\frac{d}{dt}(g_{\nu\beta}\dot{x}^\beta). \tag{7.13}$$

We shall use this in the next section. The expanded form of this equation is

$$m(g_{\nu\beta}\ddot{x}^\beta) + m\frac{\partial g_{\nu\beta}}{\partial x^\alpha}\dot{x}^\alpha\dot{x}^\beta = \frac{1}{2}m\frac{\partial g_{\alpha\beta}}{\partial x^\nu}\dot{x}^\alpha\dot{x}^\beta. \tag{7.14}$$

We notice, and it is not surprising, that the mass cancels off identically:

The free motion of a particle in a curved space is independent of the mass of the particle. The trajectory only depends on the initial conditions.

One obtains the equation for the $\ddot{x}^\mu$ by multiplying (7.14) by $g^{\mu\nu}$ (7.2),

$$\ddot{x}^\mu + \Gamma^\mu_{\alpha,\beta}\dot{x}^\alpha\dot{x}^\beta = 0, \tag{7.15}$$

where we follow the usual conventions of general relativity by introducing the *Christoffel symbols* $\Gamma^\mu_{\alpha,\beta}$ defined by

$$\Gamma^\mu_{\alpha,\beta} = \frac{1}{2}g^{\mu\nu}\left(\frac{\partial g_{\alpha\nu}}{\partial x^\beta} + \frac{\partial g_{\beta\nu}}{\partial x^\alpha} - \frac{\partial g_{\alpha\beta}}{\partial x^\nu}\right). \tag{7.16}$$

We shall make no further use of these symbols in this chapter, but we mention them in the next. It is a good example to show that the formal complexity of general relativity is a matter of writing; what is subtle is the physics.

7.3.3 Simple Examples

1. **Motion on S^2**

   One can, as an exercise, recover that the motion on a usual sphere S^2 is a uniform motion on a great circle.

2. **Free motion on S^3**

   Consider now the case of the three-dimensional "spherical" space of (7.3); i.e., the free motion on the sphere S^3. Obviously, the volume of this space is also finite since $\rho^2 = x^2 + y^2 + z^2 \leq R^2$.

   The motion exhibits more diversity that on S^2.

   In spherical coordinates, the Lagrangian of the problem is

$$\mathcal{L} = \frac{m}{2}\left(\dot{\rho}^2\left(\frac{R^2}{R^2 - \rho^2}\right) + \rho^2\dot{\theta}^2 + \rho^2\sin^2\theta\dot{\phi}^2\right). \tag{7.17}$$

   This is fully treated in problem 7.8 below. It is useful to do that exercise. We nevertheless give the main results.

   The conservation laws of the problem which bring simplifications to the motion are

   (a) There is rotational invariance. The angular momentum is conserved, and the motion occurs on a plane. Therefore we can choose the direction of the angular momentum as polar axis; i.e., $\theta = \pi/2$ and $\dot{\theta} = 0$.

   (b) The Lagrangian of the planar motion therefore reduces to

$$\mathcal{L} = \frac{m}{2} \left(\dot{\rho}^2 \left(\frac{R^2}{R^2 - \rho^2} \right) + \rho^2 \dot{\phi}^2 \right). \tag{7.18}$$

(c) The conservation of angular momentum, second Kepler's law, results in

$$\frac{d}{dt}(\rho^2 \dot{\phi}) = 0 \implies \dot{\phi} = \frac{A}{\rho^2}, \tag{7.19}$$

where A is a constant, fixed by the initial conditions.

(d) The energy, which is a constant of the motion, is therefore

$$E = \frac{m}{2} \left(\dot{\rho}^2 \frac{R^2}{R^2 - \rho^2} + \frac{A^2}{\rho^2} \right). \tag{7.20}$$

The two constants of the motion E and A satisfy the inequality

$$A^2 \leq \frac{2R^2 E}{m}, \tag{7.21}$$

which is a direct consequence of the fact that the energy is greater than the rotational energy $mA^2/2\rho^2$. This is a consequence of (7.135); i.e., $E \geq mA^2/2\rho^2 \geq mA^2/2R^2$.

The solution is simple. We define parameters ω and γ by

$$\omega^2 = \frac{2E}{mR^2} \quad \text{and} \quad \gamma^2 = \frac{mA^2}{2ER^2}. \tag{7.22}$$

From (7.136), we have the inequality

$$\gamma^2 \leq 1. \tag{7.23}$$

We conclude the following.

(a) Consider the *Euclidean* plane of the motion (i.e., $x = \rho \cos \phi, \ y = \rho \sin \phi$). For simplicity, we choose the initial parameters as $t_0 = 0, \phi_0 = 0$, and we have
$$x = R \cos \omega t \qquad y = \gamma R \sin \omega t.$$

The trajectory is an ellipse of equation $x^2 + y^2/\gamma^2 = R^2$.

(b) The point $\rho = R$ (i.e., the boundary of the space) is *always* reached, whatever the initial conditions on the energy and the angular momentum. If $A = 0$, the angular momentum vanishes and the motion is linear and sinusoidal. If $\gamma = 1$, the motion is uniform on a circle of radius R.

(c) In this Euclidean plane, the energy of the particle is

$$E = \frac{1}{2}m(\dot{x}^2 + \dot{y}^2) + V,$$

where the "effective potential" V is energy dependent:

$$V = \frac{1}{2}m\frac{\rho^2\dot{\rho}^2}{R^2 - \rho^2} = \frac{E\rho^2}{R^2}.$$

Therefore, the motion *also* appears as a two-dimensional harmonic motion whose frequency Ω depends on the *total energy E*,

$$\Omega^2 = \frac{2E}{mR^2}.$$

(d) Of course, if the square of the velocity is a constant in the *curved* four-dimensional space, this is not the case if one visualizes the phenomenon in a Euclidean plane, as above.

(e) The simplicity of the result is intuitive. Quite obviously, as one can see in the definition (7.3), the symmetry of the problem is much larger than the sole rotation in $\mathcal{R}^3$. There is a rotation invariance in $\mathcal{R}^4$. The solutions of maximal symmetry correspond to a uniform motion on a circle of radius R in a plane whose orientation is arbitrary in $\mathcal{R}^4$. The whole set of solutions is obtained by projecting these particular solutions on planes of $\mathcal{R}^3$, which leads to the elliptic trajectories that we have found.

7.4 Geodesic Lines

The following property is fundamental.

Theorem 4 *The trajectories of a free particle in a curved space are the geodesic lines of this space.*

7.4.1 Definition

In the case of a positive metric, a possible definition of a geodesic line, hereafter called a geodesic for simplicity, going through two points A and B is that it is the curve of minimal (extremal) length between these two points. In differential geometry, there are other equivalent definitions.[5]

Therefore, a geodesic is the path that minimizes the length

---

[5] In Minkowski space, light rays follow trajectories of vanishing "length." The notion of *parallel transport* allows us to overcome this apparent difficulty; see [8, 21], or [23].

$$s_{AB} = \int_A^B ds = \int_A^B \sqrt{g_{\alpha\beta} dx^\alpha \, dx^\beta}; \tag{7.24}$$

i.e., the path $\{X^\alpha\}$ such that $\delta s_{AB} = 0$ for any infinitesimal variation $\{\delta x^\alpha, \delta \dot{x}^\alpha\}$. Considering an arbitrary parameterization $\{x^\alpha(\lambda)\}$ of the path, we must find the path that minimizes the integral

$$s_{AB} = \int_A^B ds = \int_A^B \sqrt{g_{\alpha\beta} \frac{dx^\alpha}{d\lambda} \frac{dx^\beta}{d\lambda}} \, d\lambda. \tag{7.25}$$

The assertion above is that these paths are the same as those along which the action

$$S = \int_A^B \mathcal{L} \, dt = \frac{1}{2} m \int_A^B g_{\alpha\beta} \frac{dx^\alpha}{dt} \frac{dx^\beta}{dt} \, dt \tag{7.26}$$

is stationary.

7.4.2   Equation of the Geodesics

The variational problem posed in Eq. (7.25) is similar in every way with those considered in Chap. 2. Consider a variation

$$x^\nu \to x^\nu + \epsilon^\nu \quad \text{and} \quad \dot{x}^\nu \to \dot{x}^\nu + \dot{\epsilon}^\nu, \tag{7.27}$$

where

$$\dot{\epsilon}^\nu = \frac{d\epsilon^\nu}{d\lambda} \quad \text{and} \quad \epsilon^\nu(A) = \epsilon^\nu(B) = 0. \tag{7.28}$$

To first order, the variation of s_{AB} is

$$\delta s_{AB} = \int_A^B \frac{1}{2\sqrt{g_{\alpha\beta} \dot{x}^\alpha \dot{x}^\beta}} \left(\frac{\partial g_{\alpha\beta}}{\partial x^\nu} \dot{x}^\alpha \dot{x}^\beta \epsilon^\nu + 2 g_{\nu\beta} \dot{x}^\beta \dot{\epsilon}^\nu \right) d\lambda, \tag{7.29}$$

where we have set $\dot{x}^\alpha \equiv (d/d\lambda) x^\alpha$. We now integrate the second term by parts. Consider the quantity

$$F = \sqrt{g_{\alpha\beta} \dot{x}^\alpha \dot{x}^\beta}.$$

We have

$$\delta s_{AB} = \int_A^B \left[\frac{1}{2F} \left(\frac{\partial g_{\alpha\beta}}{\partial x^\nu} \dot{x}^\alpha \dot{x}^\beta - \frac{d}{d\lambda}(2 g_{\nu\beta} \dot{x}^\beta) \right) - (2 g_{\nu\beta} \dot{x}^\beta) \frac{d}{d\lambda} \left(\frac{1}{2F} \right) \right] \epsilon^\nu d\lambda = 0. \tag{7.30}$$

This variation must vanish for any $\{\epsilon^\nu\}$, and we obtain the equations

$$\frac{1}{2F}\left(\frac{\partial g_{\alpha\beta}}{\partial x^\nu}\dot{x}^\alpha\dot{x}^\beta - \frac{d}{d\lambda}(2g_{\nu\beta}\dot{x}^\beta)\right) - (2g_{\nu\beta}\dot{x}^\beta)\frac{d}{d\lambda}\left(\frac{1}{2F}\right) = 0 \quad \forall\nu. \qquad (7.31)$$

These equations are simplified if one makes an appropriate choice of the parameter λ. Consider the choice $\lambda = s$; i.e., λ is the length along the *geodesic* and $d\lambda = ds$.[6] Then, by definition, inserting this in Eq. (7.25), we have *along the geodesic*

$$F = 1, \quad \text{and} \quad \frac{dF}{d\lambda} = 0.$$

Consequently, the equation of the geodesic becomes

$$\left(\frac{\partial g_{\alpha\beta}}{\partial x^\nu}\frac{\partial x^\alpha}{\partial s}\frac{\partial x^\beta}{\partial s} - \frac{d}{ds}\left(2g_{\nu\beta}\frac{\partial x^\beta}{\partial s}\right)\right) = 0 \quad \forall\nu. \qquad (7.32)$$

Not only does this equation have the same form as the equation of motion (7.13), but it is *equivalent* to it. Indeed, we can choose $\lambda = t$. In the case of a free motion, we have seen that $v = ds/dt$ is a constant along the trajectory. Therefore, $ds = vdt$ and the factor $1/v^2$ cancels off identically in (7.32).

We have proven our assertion. The trajectories of a free particle in a curved space are the geodesics of the space. In other words, the trajectory followed by a free particle to go from A to B in a curved space is the *path of shortest length*.[7] Galileo's principle of inertia appears as a particular case of this property in flat space.

7.4.3 Examples

If we keep in mind how we have treated Example 2 of the free motion on $\mathbf{S}^3$, we can use the constants of the motion in order to determine the geodesics in simple but non trivial cases that are not totally academic.

1. **Isotropic spaces**
   Metrics of the form (7.7), or more generally

$$ds^2 = g(\rho)d\rho^2 + \rho^2 d\theta^2 + \rho^2 \sin^2\theta d\phi^2, \qquad (7.33)$$

   are isotropic and time independent. Therefore, as we have seen in Example 2, the conservation laws of energy and angular momentum simplify considerably the determination of the geodesics.

---

[6] Actually, it suffices that λ be an affine function of s.

[7] As already mentioned, in full generality this statement is wrong. This will be important when time is included. In Minkowski space there exist nontrivial zero-length paths; i.e., photon trajectories. The assertion becomes correct and general if one uses the notion of parallel transport.

Starting from (7.33) and the conservation of angular momentum, which if we choose the polar axis perpendicular to the trajectory is expressed as $\dot{\phi} = A/\rho^2$, one obtains

$$\frac{2E}{m} = g(\rho)\dot{\rho}^2 + \frac{A^2}{\rho^2} \tag{7.34}$$

and, by a simple quadrature, the relation between t and ρ,

$$\int_{\rho_0}^{\rho} d\sigma \sqrt{\left(\frac{g(\sigma)}{(2E/m - A^2/\sigma^2)}\right)} = t - t_0. \tag{7.35}$$

The only difficulty lies in the inversion of this formula in order to obtain the dependence $\rho(t)$.

2. **Hyperbolic geodesics**
   Consider the metric

   $$ds^2 = \frac{R^2}{\rho^2 + R^2} \, d\rho^2 + \rho^2 \, d\theta^2 + \rho^2 \sin^2 \theta \, d\phi^2, \tag{7.36}$$

   where R is a characteristic length.
   Notice that one considers this metric as deriving from a "Lorentzian" metric

   $$ds^2 = dx^2 + dy^2 + dz^2 - dw^2, \tag{7.37}$$

   by the three-dimensional reduction

   $$w^2 = x^2 + y^2 + z^2 + R^2 = \rho^2 + R^2. \tag{7.38}$$

   The calculation is very similar to that of example 2.
   In spherical coordinates, the Lagrangian of the problem is

   $$\mathcal{L} = \frac{m}{2} \left(\dot{\rho}^2 \left(\frac{R^2}{\rho^2 + R^2} \right) + \rho^2 \dot{\theta}^2 + \rho^2 \sin^2 \theta \dot{\phi}^2 \right). \tag{7.39}$$

   The conservation laws are the same as before.

   (a) Rotation invariance yields the conservation of angular momentum. The motion is planar.
   (b) Choosing the direction of the angular momentum as the polar axis, we have $\theta = \pi/2$ and $\dot{\theta} = 0$.
   (c) The Lagrangian of the planar motion reduces to

   $$\mathcal{L} = \frac{m}{2} \left(\dot{\rho}^2 \left(\frac{R^2}{\rho^2 + R^2} \right) + \rho^2 \dot{\phi}^2 \right). \tag{7.40}$$

(d) The conservation of angular momentum leads to

$$\frac{d}{dt}(\rho^2 \dot{\phi}) = 0 \quad \Longrightarrow \quad \dot{\phi} = \frac{A}{\rho^2} \tag{7.41}$$

where A is a constant of the motion.

(e) The energy, which is a constant of the motion, is

$$E = \frac{m}{2}\left(\dot{\rho}^2 \frac{R^2}{\rho^2 + R^2} + \frac{A^2}{\rho^2}\right). \tag{7.42}$$

The solution of the problem is obtained rather easily. One defines the parameters ω and γ as before,

$$\omega^2 = \frac{2E}{mR^2} \quad \text{and} \quad \gamma^2 = \frac{mA^2}{2ER^2}. \tag{7.43}$$

We obtain in a way similar to example 2, except that hyperbolic functions replace trigonometric functions,

$$\rho(t) = R\sqrt{\gamma^2 \cosh^2 \omega(t - t_0) + \sinh^2 \omega(t - t_0)} \tag{7.44}$$

and

$$\tan(\phi(t) - \phi_0) = \gamma \coth \omega(t - t_0). \tag{7.45}$$

We notice that the distance to the origin increases *exponentially* when $|t| \to \infty$. The geodesics of the metric (7.126) are hyperbolas

$$x^2 - y^2/\gamma^2 = R^2. \tag{7.46}$$

7.4.4 Maupertuis Principle and Geodesics

Consider again a conservative Lagrangian problem (i.e., a problem where the Lagrangian does not depend on time and the energy E is conserved, as in Chap. 4 Sect. 5.2.4). We want to show that the motion of a particle in a field of forces in Euclidean space can be reduced to the *free* motion of this particle in a curved space.

The Maupertuis principle can be extended to an arbitrary number N of degrees of freedom with $N = 3k$ for k particles in three-dimensional space. We denote the coordinates by $q_i, i = 1, \ldots, N$ and their time derivatives by $\dot{q}_i$. For simplicity, we call q the set $q_i, i = 1, \ldots, N$.

Consider in three-dimensional Euclidean space a Lagrangian of the form

$$\mathcal{L} = \frac{1}{2} \sum_{i,j=1}^{N} m_{i,j}(q)\dot{q}_i\dot{q}_j - V(q). \tag{7.47}$$

Here, we denote by $m_{i,j}(q)$ the coefficients of the quadratic form that constitutes the kinetic energy. In Cartesian coordinates, $m_{i,j}(q)$ is diagonal and does not depend on the coordinates. This is no longer true in general.

The conjugate momenta are

$$p_i = \frac{\partial \mathcal{L}}{\partial \dot{q}_i} = \sum_{j=1}^{N} m_{i,j}(q)\dot{q}_j. \tag{7.48}$$

The energy, which is a constant of the motion, is

$$E = \frac{1}{2} \sum_{i,j=1}^{N} m_{i,j}(q)\dot{q}_i\dot{q}_j + V(q). \tag{7.49}$$

Consider the time interval dt to go from point q_i, $i = 1, \ldots, N$ to point $q_i + d\,q_i$, $i = 1, \ldots, N$. This interval dt is obtained easily with the previous expression (7.49) as

$$dt = \sqrt{\frac{\sum_{i,j} m_{i,j}(q)d\,q_i d\,q_j}{2(E - V(q))}}. \tag{7.50}$$

If we insert this into the expression (5.29) of the reduced action

$$d\,S_0 = \sum_{i=1}^{N} p_i\,dq_i = \sum_{j=1}^{N} m_{i,j}(q)\frac{d\,q_j}{dt}d\,q_i, \tag{7.51}$$

the reduced action takes the form

$$S_0 = \int \sqrt{2(E - V(q)) \sum_{i,j} m_{i,j}(q)d\,q_i d\,q_j}. \tag{7.52}$$

In this form, we see that the trajectory that minimizes the reduced action S_0 is a *geodesic* of a curved space whose metric, which depends on E, is given by

$$ds^2 = 2(E - V(q)) \sum_{i,j=1}^{N} m_{i,j}(q)d\,q_i d\,q_j. \tag{7.53}$$

In this form, the Maupertuis principle appears as a purely geometric statement.

Once this geodesic is determined, the motion is determined by integrating (7.50); i.e.,

$$t - t_0 = \int_{q_0}^{q} \sqrt{\frac{\sum_{i,j} m_{i,j}(q') d\, q_i' d\, q_j'}{2(E - V(q'))}}. \tag{7.54}$$

7.5 Gravitation and the Curvature of Space-Time

The scheme we have studied up to now is appealing since the equality between the inertial and the gravitational masses follows automatically. However, the theory has an embarrassing by-product in that the norm of the velocity is a constant in time.

In order to get rid of this defect, we must introduce the time variable in the problem and *extend it to space-time and not only space.*

7.5.1 Newtonian Gravitation and Relativity

The "flat" space of special relativity corresponds to the (non-Euclidean) metric[8]

$$ds^2 = c^2 dt^2 - d\mathbf{r}^2, \quad \text{i.e.} \quad g_{00} = 1, \; g_{11} = g_{22} = g_{33} = -1. \tag{7.55}$$

Our purpose here is not to enter the domain of relativistic gravitation and general relativity as a whole (see, for instance, [2, 20, 21], or [22]). We only want to introduce a curvature of space-time that, at least to lowest order in v^2/c^2, v being a characteristic velocity of the problem, allows us to recover Newton's usual equations while maintaining the nice properties encountered above, in particular the fact that the mass drops out from the equations of motion.

What metric of space-time can we choose in order to achieve this program?

We have seen in Chap. 3 that the Lagrangian of a free relativistic particle is

$$\mathcal{L} = -mc^2 \sqrt{1 - \frac{v^2}{c^2}}. \tag{7.56}$$

The Lorentz-invariant action is

$$S = -mc^2 \int_{t_1}^{t_2} \sqrt{1 - \frac{v^2}{c^2}}\, dt. \tag{7.57}$$

In the nonrelativistic limit of small velocities, the Lagrangian (7.56) is expanded up to first order as

$$\mathcal{L} = -mc^2 + \frac{1}{2}mv^2. \tag{7.58}$$

---

[8] Unfortunately, we follow the *particle-physics tradition* of having a metric with negative space components instead of positive ones as we have used above.

In nonrelativistic physics, the Lagrangian of a particle in a gravitational potential ϕ has the form

$$\mathcal{L} = \frac{1}{2}mv^2 - m\phi, \tag{7.59}$$

and the most "natural" choice to extend Eq. (7.58) would be

$$\mathcal{L} = -mc^2 + \frac{1}{2}mv^2 - m\phi, \tag{7.60}$$

and an action

$$S = \int \mathcal{L}dt = -mc \int \left(c - \frac{v^2}{2c} + \frac{\phi}{c} \right) dt. \tag{7.61}$$

Comparing this expression with

$$S = -mc \int ds, \tag{7.62}$$

we end up quite naturally, to lowest order in v^2/c^2 and in ϕ/c^2, with the expression of the invariant element

$$ds^2 = c^2 \left(1 + \frac{2\phi}{c^2} \right) dt^2 - d\mathbf{r}^2. \tag{7.63}$$

This is the simplest, or most naive, extension of the metric (7.55) which accounts for the phenomena that interest us

$$g_{00} = \left(1 + \frac{2\phi}{c^2} \right), \quad g_{\alpha\beta} = \delta_{\alpha\beta}. \tag{7.64}$$

The equation of the geodesics obviously gives us Newton's law

$$\ddot{\mathbf{r}} = -\nabla\phi, \tag{7.65}$$

which involves neither the velocity of light, since we are in non-relativistic mechanics, nor the mass, of course.

7.5.2 The Schwarzschild Metric

In the theory of general relativity, the metric is related to the mass distribution (actually to the energy-momentum tensor) by Einstein's equations.

An exact solution of these equations was given in 1916 by Karl Schwarzschild. It is the metric generated by the static gravitational field of an isotropic mass distribution of total mass M. This metric leads to

$$ds^2 = \left(1 - \frac{r_0}{\rho}\right) c^2 dt^2 - \frac{d\rho^2}{\left(1 - \frac{r_0}{\rho}\right)} - \rho^2 (d\theta^2 + \sin^2\theta \, d\phi^2), \tag{7.66}$$

where the Schwarzschild radius r_0 is given by

$$r_0 = \frac{2GM}{c^2}. \tag{7.67}$$

One can write this formula in the more general way

$$ds^2 = \left(1 + \frac{2\Phi(\rho)}{c^2}\right) c^2 dt^2 - \frac{d\rho^2}{\left(1 + \frac{2\Phi(\rho)}{c^2}\right)} - \rho^2 (d\theta^2 + \sin^2\theta \, d\phi^2), \tag{7.68}$$

$\Phi(\rho)$ being the central Newtonian potential of the mass distribution.

In this expression, it is important to define the coordinates in a precise way. The variables (θ, ϕ) are the usual angular coordinates in the reference system centered at the origin of the mass distribution. A problem remains, however, in the definition of the radial variable ρ in the presence of a gravitational field. In the expression (7.68), the physical meaning of the coordinate ρ is that the circumference of a circle centered at the origin is equal to $2\pi\rho$. The distance between two points ρ_1 and ρ_2 in the same direction (θ, ϕ) is

$$d_{12} = \int_{\rho_1}^{\rho_2} \frac{d\rho}{\sqrt{1 - \frac{r_0}{\rho}}} > \rho_2 - \rho_1. \tag{7.69}$$

Some arbitrariness remains as far as the couple of variables (ρ, t) are concerned. Here, these variables are chosen so that there is no off-diagonal term $d\rho \, dt$ in the metric.

The proof of this formula is, of course, beyond the scope of this book. One can refer to Landau and Lifshitz [2], Sect. 97, and to Misner, Thorne, and Wheeler [23], Chap. 25. One can find the complete description of black holes (i.e. physics inside the Schwarzschild radius) in [23].

The "naive" metric (7.63) is the approximation of (7.66) to lowest order in v^2/c^2 and ϕ/c^2.

We remark on the form (7.66) that its spatial part is not locally Euclidean. There is no local rotation invariance, which is intuitive since the radial variable plays a special role. When fields are weak (i.e. $r_0/\rho \ll 1$), or at large distances, one can use locally Euclidean space variables (x, y, z), and, to a good approximation, the Schwarzschild metric (7.66) is of the form

$$ds^2 = \left(1 - \frac{r_0}{r}\right) c^2 dt^2 - \left(1 + \frac{r_0}{r}\right)(dx^2 + dy^2 + dz^2)$$
$$= \left(1 - \frac{r_0}{r}\right) c^2 dt^2 - \left(1 + \frac{r_0}{r}\right)(dr^2 + r^2(d\theta^2 + \sin^2\theta\, d\phi^2)), \quad (7.70)$$

where (r, θ, ϕ) are the usual spherical coordinates. (The proof of this result can be found in [2], Sect. 97).

7.5.3   Gravitation and Time Flow

We notice that if the metric (7.64) gives us the classical Newton equation, it "curves" time at each point in space. In that respect, it is in full agreement with the general solution of Schwarzschild, which predicts a *dilation* of the proper time in an algebraically increasing gravitational potential

$$d\tau = \sqrt{g_{00}}\, dt \simeq \left(1 + \frac{\Phi}{c^2}\right) dt. \quad (7.71)$$

This effect, as well as the "twin effect" of special relativity, has been measured with great accuracy by R.F.C. Vessot and collaborators.[9] A hydrogen maser was sent to an altitude of 10,000 km by a Scout rocket, and the variation in time of its frequency was made as the gravitational potential increased (algebraically). There are many corrections, in particular due to the Doppler effect of the spacecraft and to the Earth's rotation. It was possible to test the predictions of general relativity on the variation of the pace of a clock as a function of the gravitational field with a relative accuracy of 7×10^{-5}. This was done by comparison with atomic clocks, or masers, on Earth. Up to now, it has been one of the best verifications of general relativity. The recording of the beats between the embarked maser and a test maser on Earth is shown in Fig. 7.1. (These are actually beats between signals, which are first recorded and then treated in order to take into account all physical corrections.)

7.5.4   Precession of Mercury's Perihelion

To next order, the Schwarzschild metric curves space. This causes a variety of observable phenomena in celestial mechanics. Among these is the famous precession of the perihelion of planets and comets .

Here we choose to work with the form (7.70). In fact, the value of Schwarzschild's radius is $r_s = 2GM/c^2$, $r_s = 3$ km for the sun and $r_s = 0.44$ cm for the Earth. It is

---

[9] R. F. C. Vessot, M. W. Levine, E. M. Mattison, E. L. Blomberg, T. E. Hoffman, G. U. Nystrom, B. F. Farrel, R. Decher, P. B. Eby, C. R. Baugher, J. W. Watts, D. L. Teuber, and F. D. Wills, "Test of Relativistic Gravitation with a Space-Borne Hydrogen Maser", Phys. Rev. Lett. **45**, 2081, (1980).

Fig. 7.1 Beats between a maser onboard the spacecraft launched by a Scout rocket and a maser on Earth at various instants in GMT. **a** Signal of the dipole antenna; the pointer shows the delicate moment when the spacecraft separated from the rocket (it was important that the maser onboard had not been damaged by vibrations during takeoff). During this first phase, the special relativity effect due to the velocity is dominant. **b** Time interval of "zero beat" during ascent when the velocity effect and the gravitational effect, of opposite signs, cancel each other. **c** Beat at the apogee, entirely due to the gravitational effect of general relativity. Its frequency is 0.9 Hz. **d** Zero beat at descent. **e** End of the experiment. The spacecraft enters the atmosphere and the maser onboard ceases to work (Courtesy of R.F.C. Vessot)

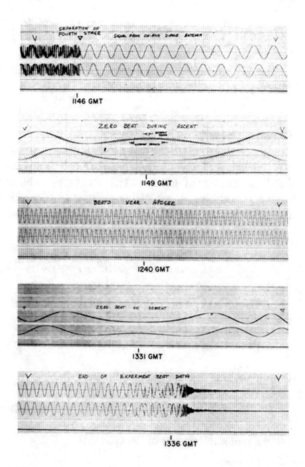

very small compared to the orders of magnitude of celestial mechanics in the solar system (1 A.U. $= 150 \times 10^6$ km). The effects are small corrections to the Newtonian terms.

The length element is given by

$$ds^2 = (1 - 2\alpha)c^2\,dt^2 - (1 + 2\alpha)(dr^2 + r^2(d\theta^2 + \sin^2\theta\,d\phi^2)), \qquad (7.72)$$

$$\text{with} \quad \alpha = \frac{GM}{rc^2} \ll 1.$$

For non-relativistic velocities, one has to good approximation

$$ds = \left((1 - \alpha)c - \frac{(1 + 3\alpha)}{2c}\left[dr^2 + r^2(d\theta^2 + \sin^2\theta\,d\phi^2)\right] \right) dt \qquad (7.73)$$

and the Lagrangian

$$\mathcal{L} = -mc\frac{ds}{dt} = -mc^2 + \frac{GMm}{r} + \frac{m}{2}\left(1 + \frac{3GM}{rc^2}\right)\left[\dot{r}^2 + r^2(\dot{\theta}^2 + \sin^2\theta \, \dot{\phi}^2)\right].$$

$$(7.74)$$

The first and most famous application is the calculation of the precession of Mercury's perihelion.

Classical Calculation

It is convenient to recall the treatment of Kepler's problem that corresponds to $\alpha = 0$. Let M be the mass of the sun (which we assume to be fixed for simplicity) and m the mass of the orbiting planet. We choose spherical coordinates (r, θ, ϕ), and we assume that the trajectory is in the plane $\theta = \pi/2$ (conservation of angular momentum). The Lagrangian of the problem is

$$\mathcal{L} = \frac{1}{2}m(\dot{r}^2 + r^2\dot{\phi}^2) + \frac{GMm}{r}.$$

Let $\mathcal{E}$ be the energy of the planet and Λ the norm of its angular momentum. For convenience, we define

$$E \equiv \frac{\mathcal{E}}{m} \quad \text{and} \quad L \equiv \frac{\Lambda}{m}. \qquad (7.75)$$

Conservation of angular momentum yields

$$r^2\dot{\phi} = L \quad \text{constant of the motion}, \qquad (7.76)$$

and the (constant) energy of the system is

$$E = \frac{1}{2}\left(\dot{r}^2 + \frac{L^2}{r^2}\right) - \frac{GM}{r}. \qquad (7.77)$$

We study the trajectory $r(\phi)$, from which the time dependence follows by using (7.76). We define $r' \equiv (dr/d\phi)$, therefore $\dot{r} = r' \, d\phi/dt = r'L/r^2$. By introducing the variable $u(\phi) = 1/r(\phi)$, we obtain the first-order equation

$$\frac{2E}{L^2} = u'^2 + u^2 - \frac{2GMu}{L^2}. \qquad (7.78)$$

The trajectory is obtained by a simple quadrature (one can take the derivative of (7.78) with respect to ϕ, which leads to a linear equation whose general solution is inserted into (7.78) in order to fix the constants):

$$u = \frac{1 + e\cos\phi}{p} \quad \text{with} \quad p = \frac{L^2}{GM} \quad \text{and} \quad e = \sqrt{1 + \frac{2EL^2}{G^2M^2}}. \qquad (7.79)$$

This also amounts to

$$r = \frac{p}{1 + e \cos \phi},$$ (7.80)

where one recognizes, for negative energies ($E \leq 0$), an ellipse of parameter p and eccentricity e.

Relativistic Correction

With the curvature of space-time, the motion remains planar. One chooses as above $\theta = \pi/2$, and the Lagrangian is given by (7.74); i.e.,

$$\mathcal{L} = \frac{1}{2}m \left(1 + \frac{3GM}{rc^2}\right)(\dot{r}^2 + r^2\dot{\phi}^2) + \frac{GMm}{r}.$$ (7.81)

There is conservation of the angular momentum $\Lambda = mr^2\dot{\phi}(1 + 3GM/(rc^2))$ and of the energy $\mathcal{E}$. As above, we set

$$E \equiv \frac{\mathcal{E}}{m}, \qquad L \equiv \frac{\Lambda}{m}, \qquad u = \frac{1}{r}.$$ (7.82)

We make use of the parameter p, the eccentricity e of the Newtonian ellipse, and the parameter λ defined by

$$p = \frac{L^2}{GM}, \qquad e = \sqrt{1 + \frac{2EL^2}{G^2M^2}}, \qquad \text{and} \qquad \lambda = \frac{3GM}{pc^2} = \frac{3G^2M^2}{L^2c^2}.$$ (7.83)

The energy is calculated as in Sect. 3.3. Its value is

$$E = \frac{1}{2}\left(1 + \frac{3GM}{c^2r}\right)(\dot{r}^2 + r^2\dot{\phi}^2) - \frac{GM}{r}.$$ (7.84)

We still define $r' \equiv (dr/d\phi)$ and $\dot{r} = r'\dot{\phi}$ so that the energy is expressed, in terms of the variables and parameters defined in (7.82) and (7.83), as

$$\frac{2E}{L^2} = \frac{(u'^2 + u^2)}{(1 + \frac{3GM}{c^2r})} - \frac{2}{p}u.$$ (7.85)

Under the change of function

$$v(\phi) = p\, u(\phi),$$ (7.86)

and multiplying by p^2, we obtain

$$(e^2 - 1) \simeq (v'^2 + v^2)(1 - \lambda v) - 2v.$$ (7.87)

Of course, we notice that in the absence of the relativistic correction ($\lambda = 0$), the solution is

$$v_0 = 1 + e \cos \phi. \tag{7.88}$$

In order to calculate the relativistic correction, we start by taking the derivative of (7.87) with respect to ϕ. We obtain

$$2v'v'' + 2v'v - 2v' - \lambda v'(3v^2 + v'^2 + 2vv'') = 0;$$

$$\text{i.e.,} \quad 2v'' + 2v - 2 - \lambda(3v^2 + v'^2 + 2vv'') = 0. \tag{7.89}$$

This is a necessary condition (the complete solution is obtained by inserting this into (7.87)).

First-Order Perturbation

The solutions of equation (7.87) can be expressed in terms of elliptic functions. However, a first-order perturbative calculation suffices since the effect is very weak. Since we have $\lambda = 3GM/pc^2 \ll 1$, we expand the solution as

$$v = v_0 + \lambda v_1 + O(\lambda^2), \tag{7.90}$$

where v_0 is the Kepler solution $v_0 = 1 + e \cos \phi$ and v_1 is the correction that interests us. Inserting this into (7.89) and retaining only the first-order terms in λ, we obtain the equation

$$v_1'' + v_1 = \frac{3 + e^2}{2} + 2e \cos \phi, \tag{7.91}$$

whose solution is

$$v_1 = \frac{3 + e^2}{2} + e\phi \sin \phi + \left(\alpha + \frac{e}{2}\right) \sin \phi, \tag{7.92}$$

α being an arbitrary constant that we choose to be equal to zero. We notice that to first order in λ the initial equation (7.87) is satisfied.

The complete solution of the problem in first-order perturbation theory, taking into account that $\cos(1 - \varepsilon)\phi \simeq \cos \phi + \varepsilon\phi \sin \phi$, is therefore

$$\frac{1}{r} = \frac{GM}{L^2} \left[1 + e \cos \left(1 - \frac{3G^2 M^2}{c^2 L^2} \right) \phi + \frac{3G^2 M^2}{c^2 L^2} \left(\frac{3 + e^2}{2} + \frac{e}{2} \sin \phi \right) \right]. \tag{7.93}$$

This is the equation of a deformed ellipse that precesses. The precession of the major axis in one period ($\delta\phi = 2\pi$) corresponds to an angle

$$\Delta\omega \simeq \frac{6\pi G^2 M^2}{c^2 L^2} \simeq \frac{6\pi G M}{c^2 a(1 - e^2)}, \qquad (7.94)$$

where a is half of the major axis of the ellipse and e its eccentricity.

The parameters of the planet Mercury are $a = 55, 3 \times 10^6$ km, $\nu = 1/T = 415$ revolutions per century, and its eccentricity is $e = 0.2056$ (the mass of the sun is $M_\odot = 2 \times 10^{30}$ kg). The calculated value is

$$\Delta\omega = 43.03 \quad \text{seconds of arc per century}$$

compared with the observed $43.11 \pm 0.45''$ per century. Einstein said that this result was the strongest emotional experience of his scientific life.[10]

7.5.5 Gravitational Deflection of Light Rays

Another effect of the metric (7.66) and the corresponding geodesics is the deviation of light rays by a gravitational field. This effect, which was one of the first verifications of general relativity, in 1919, has regained considerable interest in recent years because of its astrophysical and cosmological consequences through the gravitational lensing effect that it induces.[11] We use the weak-field approximation

$$ds^2 \sim c^2 dt^2 \left[1 + \frac{2\Phi(r)}{c^2} \right] - \left[1 - \frac{2\Phi(r)}{c^2} \right] d\mathbf{r}^2 . \qquad (7.95)$$

The most important astrophysical use of this effect is the gravitational lensing effect it produces on remote objects. This effect comes from the gravitational curvature of photon trajectories that it produces, as shown in Fig. 7.2.

In order to calculate the trajectory of a photon, we can use the fact that the proper time $d\tau$ of a photon is zero; i.e., inserting this in (7.95), we have

$$d\tau^2 = 0 = dt^2 \left[1 + \frac{2\Phi(r)}{c^2} \right] - \left[1 - \frac{2\Phi(r)}{c^2} \right] \frac{d\mathbf{r}^2}{c^2}, \qquad (7.96)$$

where $(\mathbf{r}, t)$ are the space-time coordinates of the photon as seen by an observer. From this equation, we can calculate the velocity v of a photon in a gravitational potential,

$$v = c \sqrt{\frac{1 + \frac{2\Phi(r)}{c^2}}{1 - \frac{2\Phi(r)}{c^2}}} \simeq c \left[1 + \frac{2\Phi(r)}{c^2} \right] . \qquad (7.97)$$

---

[10] See A. Pais, *Subtle is the Lord*, Chapter 14, page 253.

[11] See, for instance, J. Rich [24].

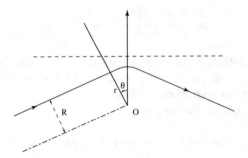

Fig. 7.2 Deviation of a photon trajectory in a gravitational potential $\Phi(r)$. This potential is assumed to be spherically symmetric. The photon position at a given time is parameterized by r and θ. The straight line represents the trajectory of a photon in the absence of a gravitational field. In the presence of this potential, the photon "falls" in the gravitational field (*full curve*)

With this expression, we can calculate the photon trajectory by the *Fermat principle* exactly as for curved rays in Eq. (2.5); i.e., by minimizing the integral

$$T = \int_A^B \frac{d\ell}{v},$$

(7.98)

where A and B are the endpoints of the photon trajectory and $d\ell$ is the length element along the trajectory.

We assume that the potential $\Phi(r)$ is spherically symmetric and centered at the origin. We consider the motion in the plane (AOB) and we use polar coordinates (r, θ) as shown in Fig. 7.2. We consider a situation where A and B are symmetric to each other, so that $\theta = 0$ corresponds to the point of shortest distance to the origin. It is convenient to determine the function $r(\theta)$ that minimizes the time T. Under these conditions, Eq. (7.98) can be written as

$$T = \frac{1}{c} \int_A^B \frac{\sqrt{1 + r^2 \dot{\theta}^2}}{[1 + \frac{2\Phi(r)}{c^2}]} dr,$$

(7.99)

where $\dot{\theta} = d\theta/dr$.

We consider the potential created by a total mass M, and we assume the photon path is outside the mass distribution so that we can set

$$\frac{2\Phi(r)}{c^2} = -\frac{2GM}{rc^2} \equiv \frac{\lambda}{r},$$

(7.100)

which defines the constant λ.

We notice in (7.99) that the variable θ is not present in the Lagrangian

$$\mathcal{L} = \frac{\sqrt{1 + r^2 \dot{\theta}^2}}{[1 + \frac{\lambda}{r}]}. \tag{7.101}$$

Therefore, the momentum $\partial \mathcal{L} / \partial \dot{\theta}$ is conserved; i.e.,

$$r^2 \dot{\theta} = R(\sqrt{1 + r^2 \dot{\theta}^2}) \left(1 + \frac{\lambda}{r}\right), \tag{7.102}$$

where R is a length that is a constant of the motion. Solving this first-order differential equation causes no special difficulty. We change to dimensionless variables

$$r = xR, \quad \theta' = d\theta/dx \equiv R\dot{\theta}, \quad \text{and} \quad \mu = \lambda/R, \tag{7.103}$$

and, by squaring and taking into account that the gravitational term is small, $\mu/x \ll 1$, we obtain

$$x^4 \theta'^2 = (1 + x^2 \theta'^2) \left(1 + \frac{2\mu}{x}\right). \tag{7.104}$$

In the absence of a gravitational field, $\mu = 0$, we obtain the equation

$$\theta' = \frac{1}{x\sqrt{x^2 - 1}}; \quad \text{i.e.,} \quad \theta = \arccos \frac{R}{r} \quad \text{or} \quad r \cos \theta = R,$$

which corresponds to a straight line at a distance R from the origin.

If we now switch on the gravitational potential, we obtain, to first order in μ/x, the equation

$$\theta' = \frac{1}{x\sqrt{x^2 - 1}} + \frac{\mu}{(\sqrt{x^2 - 1})^3}. \tag{7.105}$$

whose solution is

$$\theta = \arccos \frac{R}{r} - \frac{\lambda r}{R\sqrt{r^2 - R^2}} \tag{7.106}$$

(we have come back to the variable r).

The value of the constant R is obtained from the closest distance r_0 of the photon to the origin, which corresponds to $\theta = 0$. We obtain

$$R = r_0(1 - \varepsilon) \quad \text{with} \quad \varepsilon = \frac{GM}{r_0 c^2}.$$

One can check that, to the same order, R is nothing but the impact parameter of the photon (i.e. the distance between its trajectory, which is linear at long distances, and the parallel line going through the center $r = 0$).

What is more interesting is the angular deflection compared with a straight line. In the absence of the gravitational field, the photon follows a straight line, so that the difference between the direction of arrival and the direction of departure is $\Delta\theta_\infty^0 = \pi$. This direction of departure is also the (Euclidean) direction of observation of the source that emits the photon.

In solution (7.106), in the presence of the gravitational potential, this same difference is twice the difference $\theta(r = \infty) - \theta(r = r_0)$. By definition, $\theta(r = r_0) = 0$. For $r \rightarrow \infty$, Eq. (7.106) gives $\theta(r = \infty) = \pm\pi/2 - \lambda/R$, according to whether it is the initial or final direction of the photon. The difference between the direction of reception of the photon and the geometrical direction of its source is $\Delta\theta_\infty^{GM} = \pi - 2\lambda/R$. In other words, one observes a deflection of the light rays compared with a straight line, due to the gravitational potential, of

$$\Delta\theta_\infty = \Delta\theta_\infty^{GM} - \Delta\theta_\infty^0 = \frac{4GM}{Rc^2}, \qquad (7.107)$$

where R is the impact parameter or, to good approximation, the closest distance between the photon and the center of the potential.[12]

For a light ray coming from a star and grazing the edge of the sun, the calculated deflection is $1.75''$. In the case of Jupiter, it is $0.02''$.

The first measurement of this effect was performed by teams led by Sir Arthur Stanley Eddington.[13] It was done on the Sobral Islands in Brasil and the Principe Islands in the Gulf of Guinea on May 29, 1919. The experiment consisted in observing the apparent motion of stars (seven at Sobral and five at Principe) during a total eclipse of the sun. The results, $1.98 \pm 0.16''$ at Sobral and $1.61 \pm 0.31''$ at Principe, were in agreement with Einstein's prediction. It is most probably this experiment that generated the public's interest in relativity and the fame of Einstein himself.

The most precise measurement at present comes from interferometric radioastronomical observation of radio waves coming from the source 3C 279.[14] It gives the result $1.77 \pm 0.20''$.

7.6   Gravitational Optics and Mirages

The effect we have just described, the action of a gravitational field on the propagation of light, shows that a vast field has now been opened in what one could call *gravitational optics*.

---

[12] It is amusing that this is exactly twice as much as the deflection that a Newtonian argument would give using the Rutherford scattering formula.

[13] F.W. Dyson, A.S. Eddington, and C. Davidson, Philos. Trans. R. Soc. London, Ser. **220** A, 291 (1920); Mem. R. Astron. Soc., **62**, 291 (1920).

[14] G.A. Seielstad, R.A. Sramek, and K.W. Weller, Phys. Rev. Lett., **24**, 1373 (1970).

7.6.1 Gravitational Lensing

As we shall see, the most important cosmological use of this effect is through gravitational lensing of light on remote objects in the universe. This effect is due to the fact that mass (not only mass of galaxies but also of "dark matter") in the universe acts as an optical instrument that can enable one to observe faraway objects and therefore very "young" objects.

Two potentials are of particular interest. The first is that of a point-like mass M:

$$\frac{\partial \Phi}{\partial r} = \frac{GM}{r^2} \qquad \Rightarrow \qquad \alpha = \frac{4GM}{Rc^2} . \tag{7.108}$$

The angle of deflection α is of the order of the gravitational potential $\Phi(R)$ at the shortest distance of approach or, equivalently, the square of the velocity v_c^2/c^2 of an object orbiting on a circle at the shortest distance of approach.

The second potential of interest gives a constant rotation velocity and is a good approximation for extended objects such as galactic halos of clusters of galaxies:

$$\frac{\partial \Phi}{\partial r} = \frac{v_c^2}{r} \qquad \Rightarrow \qquad \alpha = \pi \frac{v_c^2}{c^2} . \tag{7.109}$$

Here, v_c is the (constant) circular velocity of objects orbiting in the galaxy or in the cluster of galaxies.

7.6.2 Gravitational Mirages

As shown in Fig. 7.3, the gravitational deflection can yield two images of a source. The two images have impact parameters b_1 and b_2. The potential created by a point-like mass always gives two images because the angle of deflection diverges for small values of the impact parameter. We will see later that one of the images is in general much more luminous than the other.

In the case of an extended mass distribution, such as a cluster of galaxies (7.109), one can only observe two images separated by an angle α if the undeflected impact parameter satisfies $b_0 < b_{max} = L\alpha/2$ where L is the distance between the source and the lens (which we take here to be equal to the distance between the observer and the lens for simplicity). The reason is that if b_0 were larger than $L\alpha/2$, the two images would be on the same side, which is impossible.

The large clusters correspond to $\alpha = v_c^2/c^2 \sim 10^{-5}$, and the two images can be separated by terrestrial telescopes of resolution $\sigma_\theta \sim 3 \times 10^{-6}$.

The "cross-section" necessary for a double image to occur is $\sigma \sim \pi b_{max}^2 = \pi L^2 v_c^2/c^2$. This cross-section increases with L because the necessary angle of deflection decreases with L.

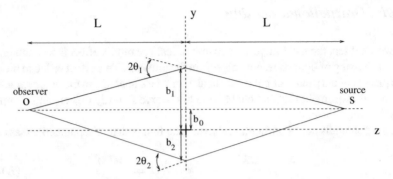

Fig. 7.3 Creation of two images of a source S by a gravitational potential symmetric around the origin. The undeflected impact parameter is b_0, whereas the physical photon trajectories have impact parameters b_1 and b_2 and deflection angles $\alpha_1 = 2\theta_1$ and $\alpha_2 = 2\theta_2$. For simplicity, the source and the observer O are assumed to be at equal distances from the origin. For clarity, the angles have been grossly exaggerated (Courtesy of James Rich)

The probability for a given object to have two images because of the lensing effect due to a cluster of galaxies is simply equal to the probability P that this object hides behind a cluster. This probability is proportional to the cross-section, to the number density n of clusters, and to the total length of the path $\sim L$:

$$P \sim 2Ln\sigma \propto L^3. \tag{7.110}$$

The fact that this probability increases rapidly with L makes the number of double-image quasars sensitive to the value of the undeflected impact parameter b_0.

A second practical application of deflection by clusters is that the time of flight is not the same for both images. Quasars have an intrinsic variability and by comparing the light curves (light flux as a function of time) one can determine the difference Δt of the two times of flight.

The time it takes light to go from one point to another can be deduced with no difficulty from the calculations performed in Sect. 7.5.5. We have

$$\frac{dr}{cdt} = \left(1 - \frac{2GM}{c^2 r}\right)\left[1 - \frac{1 - 2GM/rc^2}{1 - 2GM/Rc^2}\left(\frac{R}{r}\right)^2\right]^{1/2}. \tag{7.111}$$

To first order in $2GM/rc^2$ and $2GM/Rc^2$, the time interval to get from r to R is

$$ct(r, R) = \sqrt{r^2 - R^2} + \frac{2GM}{c^2}\ln\left(\frac{r + \sqrt{r^2 - R^2}}{R}\right) + \frac{GM}{c^2}\sqrt{\frac{r - R}{r + R}}. \tag{7.112}$$

The first term is the obvious term in the absence of a gravitational effect.

If we consider a mirage, the time delay is the difference between the integrals calculated along each path. The first-order term vanishes obviously. This leaves a "gravitational" term, which is proportional to the potential difference, and a "geometric term."

For an angle of deflection independent of the point of impact, which is approximately the case for clusters of galaxies (7.109), the geometric term vanishes, leaving

$$\Delta t \sim 2 \int_{-\infty}^{\infty} dz \, [\Phi(y_2(z)) - \Phi(y_1(z))] , \qquad (7.113)$$

where $y_1(z)$ and $y_2(z)$ are the photon trajectories in the two images. Consider the nearly symmetric case $|y_1| \sim |y_2|$. Going back to Fig. 7.3, we see that in the case where $\theta_1 \sim \theta_2 = \theta$, we have $b_1 - b_0 = b_2 + b_0$. The integral will be dominated by the nearby region of the cluster, and we can make the approximation

$$|y_1(z)| - |y_2(z)| \sim b_1 - b_2 \sim 2b_0 . \qquad (7.114)$$

Substituting this in (7.113), we obtain

$$\Delta t \sim 4b_0 \int_{-\infty}^{\infty} dz \, \frac{\partial \Phi}{\partial y} . \qquad (7.115)$$

The integral is simply the deflection angle 2θ given by (7.107), and the time delay is

$$\Delta t = 4L \, \frac{b_0}{L} \, 2\theta . \qquad (7.116)$$

The factor b_0/L is the angular separation between the center of the cluster and the average position of the two images.

In order to estimate the length of the delay, we can take $b_0/L \sim \theta \sim 10^{-5}$ and $L \sim d_{hub}$ (where d_{hub} is the Hubble distance, $d_{hub} = c/H_0 \simeq 4300 \, \text{Mpc}$, H_0 being the Hubble constant), which gives $\Delta t \sim 1$ year.

7.6.3 Observation of a Double Quasar

The first historical observation of this effect was the observation in 1979 of the "double quasar" caused by the gravitational lens Q0957+561.[15] The original image is shown in Fig. 7.4.

The two quasars have exactly the same spectrum. However, the time variation of the signals emitted is the same except for a delay of 417 days. Once the two images are subtracted from each other, taking this delay into account, the galaxy that acts as a gravitational lens appears clearly.

---

[15] D. Walsh, R.F. Carswell, and R.J. Weymann, Nature, **279**, 381 (1979).

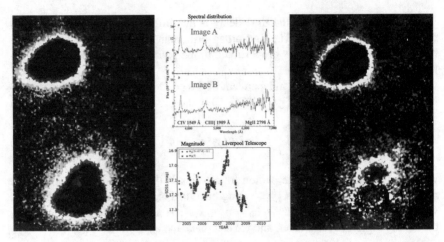

Fig. 7.4 First high resolution observations of the double quasar QSO 0957+561 A/B. *Left* image of the two spots, comparable, observed directly. Center; *Top* spectral distribution of the two spots (Photo credit D. Walsh); *Bottom* Evolution of the brightness (apparent magnitude) of the two spots over time—the *red dots* come from QSO A, with a delay of 417 days compared to the green squares of QSO B (Luis J. Goicoechea and Vyacheslav N. Shalyapin), (Photo credit: the SAO/NASA Astrophysics Data System, Liverpool Quasar Lens Monitoring (LQLM) Programme.) *Right* image obtained by subtracting the intensity of the bottom spot (QSO 0957+561 B) the intensity and spectral distribution of the QSO A (top spot), measured 417 days later. This simple subtraction, whose orientation is adjusted with precision, almost erases the image of the quasar itself, and reveals the galactic cluster G1 responsible for the mirage, with a total mass of 10,000 billion solar masses (Alan Stockton, Op. Cit.). Photo Alan Stockton

One can observe pictures with a multiplicity greater than two. The most spectacular example is perhaps the *Einstein cross* shown in Fig. 7.5. Four images of the pulsar Q2237+0305 appear, together with the spiral galaxy that causes the mirage. In the case of a straight alignment, one can observe an "Einstein ring," as shown in Fig. 7.6.

This is of course a very exceptional situation. Einstein did not believe that such a phenomenon could ever be observed. Nevertheless, he did the calculation to please his friend Mandl. In Fig. 7.6 the galaxy B1938+666 is "hidden" behind a nearby galaxy. This latter galaxy does not act as a screen but, on the contrary, as a gigantic gravitational lens. It amplifies considerably the luminosity of the first galaxy, in the shape of a nearly circular ring caused by the fact that the sun and the two galaxies are nearly exactly aligned.

In (7.116), the reflection angle 2θ can be determined from the angular separation of the two images. The angle b_0/L is more difficult to determine because it depends on the mass distribution in the cluster.

Once the two angles θ and b_0/L are determined, the distance L of the cluster can be determined by measuring the value of Δt. This determination of the distance is a very useful tool for the evaluation of Hubble's constant (see [24]).

Fig. 7.5 The Einstein cross. The four different images of the same quasar at a redshift of 1.7 are due to the central galaxy, which is at a redshift of 0.04 and therefore much closer. One can wonder about the probability of finding such an alignment, but the vastness of the universe and the perseverance of astronomers are such that events of small probability are observed in appreciable amounts (Credit NASA and ESA)

Gravitational Lens G2237+0305

Fig. 7.6 Einstein ring caused by the lensing by a close galaxy of the light emitted by the galaxy B1938+666 located behind it. The actual size of the visible object is several tens of thousands of light-years. The picture comes from the Hubble Space Telescope (Credit L. J. King (U. Manchester), NICMOS, HST, NASA)

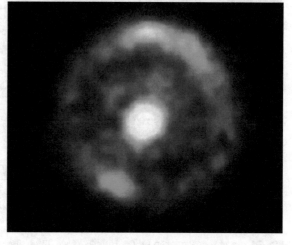

The last effect of gravitational lensing comes from the distortion of the image of an extended object. The distortion along the radial and tangent directions is illustrated in Fig. 7.7. This distortion causes the arcs that can be seen in Fig. 7.8.

This effect can be used to determine the mass of the cluster. The masses determined by this effect can be compared with the visible masses and with the masses one estimates with the virial theorem and the dispersion in velocities. It is a method to evaluate the amount of dark matter in the cluster.

In this respect, the universe appears as an endless gallery of mirages.

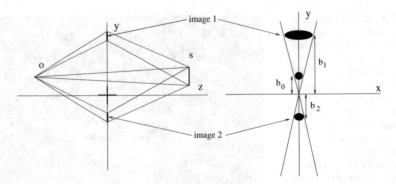

Fig. 7.7 Gravitational lensing effect by a spherically symmetric potential centered at the origin on the light emitted by an extended object S in the background of the sky. In this example, two images are seen by the observer O. The *right-hand panel* shows a projection on the x, y plane (at the z value of the lens) of the two images. The image one would observe in the absence of the lens is also shown. Owing to the cylindrical symmetry, the motion of photons is planar and the two images are therefore extended in the tangent direction. (If the object were exactly behind the lens, the image would be a circle around the origin.) Owing to this distortion, galaxies of the background of the sky can appear as arcs, as one can see in Fig. 7.8 (Figure courtesy of James Rich)

Fig. 7.8 In this image taken by the Hubble Space Telescope, virtually all of the bright objects are galaxies of the giant cluster Abell 370 surrounded by arches, giant gravitational mirages. This galaxy cluster is so massive and compact that its gravitational field focuses the light of the galaxies *behind* it. The result is images stretched over arcs, a simple focusing effect analogous to what can be observed through a glass of water looking at lights of the city at night. The Abell 370 cluster is 5 billion light-years away in the constellation of the Dragon. The spectrum of these arcs is very strongly shifted towards blue compared to that of the stars in the cluster because the focused light comes from very young and hot stars at the beginning of their evolution (Photo credits NASA, ESA/Hubble, HST Frontier Fields)

Fig. 7.9 Sketch of a gravitational lensing effect in the Large Magellanic Cloud (LMC) by an invisible object in the galactic halo. The two images cannot be resolved, but the combined luminosity of the two images gives rise to a time-dependent amplification of the light of the star when the invisible object crosses the line of sight. The corresponding light curve is shown in Fig. 7.10

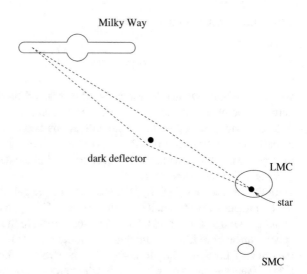

7.6.4   Baryonic Dark Matter

The theory of nucleosynthesis indicates that the baryon density in the universe is 0.04 times the critical density. This leads to the idea that baryons cannot account for all the dark matter. It is nevertheless possible that baryons are a component of the *galactic* dark matter if they are in a form that does not emit light in appreciable amounts. The simplest way this can happen is if the baryons are hidden in objects that either do not burn (for instance, brown dwarfs) or have ceased to burn   (for instance, white dwarfs, neutron stars, or black holes). Brown dwarfs have a mass $< 0.07 M_\odot$, which is not sufficient to create high enough temperatures for the burning of hydrogen. Initially, they were the preferred candidates because they completely avoid the problems associated with background light emission or the pollution of the interstellar medium by heavy elements caused by mass loss or supernova explosions.

The dark objects located in galactic halos are called "machos", for "massive compact halo objects."

Paczyński[16] has suggested that machos could be detected through the gravitational lensing effect they induce on individual stars of the Large Magellanic Cloud (LMC) (Fig. 7.9). This small galaxy is 50 kpc away from the Earth.

The theory of gravitational lensing was done above. For a point-like lens, it is simple to show that the two amplifications[17] depend on the reduced impact parameter $u = b_0 / R_\text{E}$,

$$A_\pm = \frac{u^2 + 2 \pm u\sqrt{u^2 + 4}}{2u\sqrt{u^2 + 4}} , \qquad (7.117)$$

---

[16] B. Paczyński, Astrophys. J. **304**, 1 (1986).

[17] Professionals prefer the term "magnification."

where the "Einstein radius" is given by

$$R_{\rm E}^2 = \frac{4GMLx(1-x)}{c^2} \, , \qquad (7.118)$$

where L_x is the distance between the observer and the lens, L being the distance between the observer and the source. We see that for $b_0 \gg R_{\rm E}$, $A_+ = 1$ and $A_- = 0$, as expected. For $b_0 \to 0$, the amplifications become infinite formally, and this actually results from the fact that a point-like source is deformed into a ring. The divergence ceases if one takes into account the finite extension of the source, which gives an effective extension at b_0.

In the case of "lensing" by stellar objects of the galactic halo, the angle between the two images is very small (<1 milliarcsec). This type of effect is therefore called "microlensing". Terrestrial telescopes cannot resolve the two images because atmospheric turbulence blurs the images and stellar objects have an angular dimension of the order of 1 arcsec. Therefore, the only observable effect is a temporary amplification of the total intensity when the macho comes close to the line of sight, and then recedes from it. The amplification is

$$A = \frac{u^2+2}{u\sqrt{u^2+4}} \, , \qquad (7.119)$$

where u is the closest distance to the (undeflected) line of sight that the deflector reaches in units of "Einstein's radius," $R_{\rm E} = \sqrt{4GMLx(1-x)/c^2}$, where L is the distance between the source and the observer, L_x is the distance between the observer and the deflector, and M is the mass of the macho.

The amplification is larger than 1.34 when the distance to the line of sight is less than $R_{\rm E}$. This amplification corresponds to an acceptable observational threshold since photometry can be performed quite easily to better than 10% accuracy. At a given moment, the probability P for a star to be amplified by a factor 1.34 is equal to the probability that its undeflected light path passes within one Einstein radius of the macho,

$$P \sim n_{\rm macho} \, L \, \pi R_{\rm E}^2 \, , \qquad (7.120)$$

where $n_{\rm macho}$ is the average number density of machos lying between the LMC and us, and L is the distance of the LMC. The macho density is $n_{\rm macho} \sim M_{\rm halo}/ML^3$, where $M_{\rm halo}$ is the total mass of the halo up to the position of the LMC. Using the expression of the Einstein radius, one finds that P does not depend on the mass M but is determined only by the velocity of the LMC:

$$P \sim \frac{GM_{\rm halo}}{Lc^2} \sim \frac{v_{\rm LMC}^2}{c^2} \, . \qquad (7.121)$$

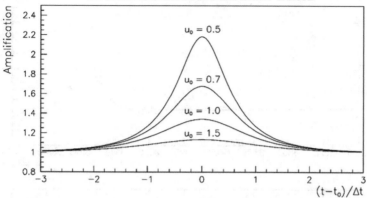

Fig. 7.10 Microlensing curves for a point-like source. The curves correspond to four values of the closest approach distance (0.5, 0.7, 1.0, and 1.5 times the Einstein radius). The duration timescale Δt, which depends on the mass, is normalized according to (7.122). (Courtesy of James Rich.)

The LMC is believed to orbit around the Milky Way with a velocity of $v_{\text{LMC}} \sim$ 200 km/s. (This corresponds to a flat rotation curve up to the LMC.) In that case, P is of the order of 10^{-6}. More refined calculations give $P = 0.5 \times 10^{-6}$ [24].

Since the observer, the star, and the deflector are in relative motion with respect to one another, a noticeable amplification lasts only as long as the non-deflected ray remains within one Einstein radius. The resulting light curve, which is achromatic and symmetric, is represented in Fig. 7.10 for a variety of values of the impact parameter. The timescale of the amplification is the time Δt that it takes the object to cross one Einstein radius between the observer and the source. For the lensing of stars of the LMC and deflectors of our halo, the relative velocities are of the order of 200 km s^{-1} and the position of the deflector is roughly halfway between the observer and the source ($x \sim 0.5$). The average Δt is then

$$\Delta t \sim \frac{R_{\text{E}}}{200 \text{ km/s}} \sim 75 \text{ days} \sqrt{\frac{M}{M_{\odot}}}. \tag{7.122}$$

The duration distribution can therefore be used to estimate the mass of machos, assuming they are in the galactic halo (and not in the LMC).

Two experimental groups, the MACHO collaboration and the EROS collaboration, have published results of searches for events in the directions of the LMC and the SMC (the nearby Small Magellanic Cloud). The absence of events lasting less than 15 days allowed both groups to exclude objects of masses in the interval $10^{-7} M_{\odot} < M < 10^{-1} M_{\odot}$ as the main component of the halo.[18] these limits exclude brown dwarfs of masses $\sim 0.07 M_{\odot}$ as the major components of the halo.

---

[18] C. Alcock, R.A. Allsman, D. Alves, et al., Astrophys. J. Lett. **499**, L9 (1998).

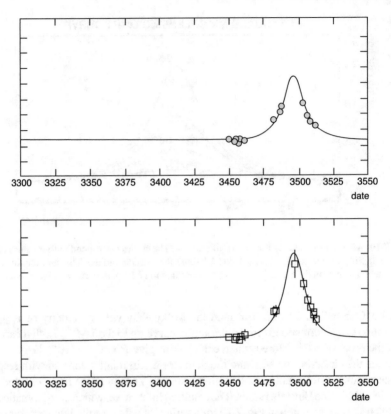

Fig. 7.11 Gravitational microlensing event in the EROS experiment. The *upper picture* is in the blue part of the optical spectrum and the *lower picture* is in the *red* part. The phenomenon is symmetric and achromatic, as expected (Courtesy of James Rich)

One important aspect of the phenomenon is that the amplification should be *symmetric* and *achromatic*. Figure 7.11 shows an event recorded by the EROS experiment that possesses both properties.

The MACHO collaboration, however, observed events of a duration of ~ 50 days.[19] If these are interpreted as originating from dark lenses of the galactic halo, this rate corresponds to a fraction $f = 0.2$ of machos contributing to the total mass of the halo. The timescale corresponds to objects of mass $\sim 0.4 M_\odot$.

EROS has only published upper bounds on the fraction of the halo components made of machos.[20]

The results of the two experiments show that it is unlikely that the halo is made up predominantly of objects of the order of stellar masses. The present challenge is to prove that the events observed by the MACHO collaboration are caused by lensing by objects of the halo and not, for instance, lensing objects in the clouds themselves.

---

[19] C. Alcock, R. Allsman, D.R. Alves, et al., Astrophys. J. **542**, 281 (2000).

[20] T. Lasserre, et al., EROS Collaboration, Astron. Astrophy. **355**, L39 (2000).

If that is the case, the mass estimation implies that they correspond to very old white dwarfs, perhaps the oldest stars.

The information on the localization of the lenses (in the galactic halo or in the Magellanic Clouds) is difficult to obtain. The simplest case is that of events with a very large amplification, in particular the events due to binary lenses. In such events, the light curve is modified in a way that depends on the relative distance of the lens and the source stars.[21] It is also possible to obtain information on the distance of lenses in events of very long duration when the motion of the Earth around the sun modifies the light curve.[22] In the future, it will also be possible to resolve the two microlensing images with interferometric space telescopes. Such measurements will give enough information to determine the distances of the lensing objects and to draw a definite conclusion.

The search for dark objects by microlensing is under way in the Andromeda Nebula, the spiral galaxy M31, which is close to our galaxy.[23]

Experimental results on these problems and on general relativity can be found in the book by James Rich [24].

7.7 Exercises

7.1. Geodesics

Consider the metric

$$ds^2 = \frac{R^2}{\rho^2 - R^2} \, d\rho^2 + \rho^2 \, d\theta^2 + \rho^2 \sin^2 \theta \, d\phi^2, \qquad (7.123)$$

which can be considered to be derived from a "Lorentzian" metric

$$ds^2 = dx^2 + dy^2 + dz^2 - dw^2, \qquad (7.124)$$

by the three-dimensional restriction

$$w^2 = x^2 + y^2 + z^2 - R^2 = \rho^2 - R^2. \qquad (7.125)$$

Find a parametric expression for the geodesics in terms of the time t.

---

[21] C. Afonso, C. Alard, J.N. Albert, et al., Astrophys. J. **532**, 340 (2000).
[22] C. Alcock, R. A. Allsman, D. Alves, et al., Astrophys. J. Lett. **454**, L125 (1995).
[23] A.P.S. Crotts and A. B. Tomaney, Astrophys. J. Lett., **473**, L87 (1995).

7.2. Hyperbolic Geodesics

Consider the metric

$$ds^2 = \frac{R^2}{\rho^2 + R^2}\, d\rho^2 + \rho^2\, d\theta^2 + \rho^2 \sin^2 \theta\, d\phi^2, \tag{7.126}$$

where R is a positive characteristic length.

  Notice that one considers this metric as deriving from a "Lorentzian" metric, if one changes the sign of R

$$ds^2 = dx^2 + dy^2 + dz^2 - dw^2, \tag{7.127}$$

by the three-dimensional reduction

$$w^2 = x^2 + y^2 + z^2 + R^2 = \rho^2 + R^2. \tag{7.128}$$

  The calculation is very similar to that of Example 2.

  In spherical coordinates, the Lagrangian of the problem is

$$\mathcal{L} = \frac{m}{2}\left(\dot{\rho}^2 \left(\frac{R^2}{\rho^2 + R^2} \right) + \rho^2 \dot{\theta}^2 + \rho^2 \sin^2 \theta \dot{\phi}^2 \right). \tag{7.129}$$

The conservation laws are the same as before.

1. Rotation invariance yields the conservation of angular momentum. The motion is planar.
2. Choosing the direction of the angular momentum as the polar axis, we have $\theta = \pi/2$ and $\dot{\theta} = 0$.
3. The Lagrangian of the planar motion reduces to

$$\mathcal{L} = \frac{m}{2}\left(\dot{\rho}^2 \left(\frac{R^2}{\rho^2 + R^2} \right) + \rho^2 \dot{\phi}^2 \right). \tag{7.130}$$

4. The conservation of angular momentum leads to

$$\frac{d}{dt}(\rho^2 \dot{\phi}) = 0 \quad \Longrightarrow \quad \dot{\phi} = \frac{A}{\rho^2} \tag{7.131}$$

  where A is a constant of the motion.

1. Calculate the energy of the system
2. Derive the parameters of the equation of motion, $\rho(t)$ and $\phi(t)$.
3. How does the distance to the origin behave as $|t| \to \infty$?
4. What are the equations of the geodesics in the (x, y) plane?

7.8 Problem. Motion on the Sphere S³

Consider the case of free motion on the three-dimensional "spherical" space of a sphere S^3 imbedded in $\mathcal{R}^4$

Obviously, the volume of this space is finite since $\rho^2 = x^2 + y^2 + z^2 \leq R^2$.

In spherical coordinates, the Lagrangian of the problem is

$$\mathcal{L} = \frac{m}{2} \left(\dot{\rho}^2 \left(\frac{R^2}{R^2 - \rho^2} \right) + \rho^2 \dot{\theta}^2 + \rho^2 \sin^2 \theta \dot{\phi}^2 \right). \tag{7.132}$$

Prove the conservation laws of the problem which bring simplifications to the motion:

1. There is rotational invariance. The angular momentum is conserved, and the motion occurs on a plane.
2. We can choose the direction of the angular momentum as polar axis; i.e., $\theta = \pi/2$ and $\dot{\theta} = 0$.
3. The Lagrangian of the planar motion therefore reduces to

$$\mathcal{L} = \frac{m}{2} \left(\dot{\rho}^2 \left(\frac{R^2}{R^2 - \rho^2} \right) + \rho^2 \dot{\phi}^2 \right). \tag{7.133}$$

4. The conservation of angular momentum, second Kepler's law, results in

$$\frac{d}{dt} (\rho^2 \dot{\phi}) = 0 \implies \dot{\phi} = \frac{A}{\rho^2}, \tag{7.134}$$

where A is a constant, fixed by the initial conditions.
5. The energy, which is a constant of the motion, is therefore

$$E = \frac{m}{2} \left(\dot{\rho}^2 \frac{R^2}{R^2 - \rho^2} + \frac{A^2}{\rho^2} \right). \tag{7.135}$$

The two constants of the motion E and A satisfy the inequality

$$A^2 \leq \frac{2R^2 E}{m}, \tag{7.136}$$

which is a direct consequence of the fact that the energy is greater than the rotational energy $mA^2/2\rho^2$. This is a consequence of (7.135); i.e., $E \geq mA^2/2\rho^2 \geq mA^2/2R^2$.

The Eqs. (7.134) and (7.135) are first-order differential equations that determine the motion in terms of the constants of the motion E and A.

The solution is simple. We define parameters ω and γ by

$$\omega^2 = \frac{2E}{mR^2} \quad \text{and} \quad \gamma^2 = \frac{mA^2}{2ER^2}. \tag{7.137}$$

From (7.136), we have the inequality

$$\gamma^2 \le 1. \tag{7.138}$$

Setting

$$\rho = R\cos(\omega\psi); \quad \text{i.e.,} \quad \dot{\rho} = -\omega\dot{\psi}R\sin(\omega\psi). \tag{7.139}$$

If we insert this in Eq. (7.135), we obtain

$$\omega^2 = \omega^2\dot{\psi}^2 + \frac{\omega^2\gamma^2}{\cos^2(\omega\psi)}; \tag{7.140}$$

i.e.,

$$\omega^2\dot{\psi}^2 \cos^2(\omega\psi) = \omega^2 (\cos^2(\omega\psi) - \gamma^2). \tag{7.141}$$

We now make the change of functions

$$\sin(\omega\psi(t)) = \sqrt{1-\gamma^2}\,u(t); \quad \text{therefore } \cos^2(\omega\psi) = 1 - (1-\gamma^2)u^2. \tag{7.142}$$

Show that the choice

$$u(t) = \sin(\omega\zeta(t)) \tag{7.143}$$

leads with no difficulty to:

$$\dot{\zeta}^2 = 1, \quad \text{namely} \quad u = \sin(\omega(t-t_0)), \tag{7.144}$$

and to the result, i.e. The expressions of $\rho(t)$, $\tan(\phi(t) - \phi_0)$ as well as the frequency ω.

$$\rho(t) = R\sqrt{\cos^2 \omega(t-t_0) + \gamma^2 \sin^2 \omega(t-t_0)}, \tag{7.145}$$

which is periodic and of frequency ω. The calculation of the time evolution of the azimuthal angle $\phi(t)$ is obtained by this expression and Eq. (7.134),

$$\dot{\phi} = \frac{A}{R^2(\cos^2 \omega(t-t_0) + \gamma^2 \sin^2 \omega(t-t_0))}; \tag{7.146}$$

7.8 Problem. Motion on the Sphere S^3

159

i.e.,

$$\tan(\phi(t) - \phi_0) = \gamma \tan \omega(t - t_0), \tag{7.147}$$

which is also periodic and of frequency ω.

We conclude the following.

1. Consider the *Euclidean* plane of the motion (i.e., $x = \rho \cos \phi$, $y = \rho \sin \phi$). For simplicity, we choose the initial parameters as $t_0 = 0$, $\phi_0 = 0$, and we have

$$x = R \cos \omega t \qquad y = \gamma R \sin \omega t.$$

   The trajectory is an ellipse of equation $x^2 + y^2/\gamma^2 = R^2$.
2. The point $\rho = R$ (i.e., the boundary of the space) is *always* reached, whatever the initial conditions on the energy and the angular momentum. If $A = 0$, the angular momentum vanishes and the motion is linear and sinusoidal. If $\gamma = 1$, the motion is uniform on a circle of radius R.
3. In this Euclidean plane, the energy of the particle is

$$E = \frac{1}{2} m(\dot{x}^2 + \dot{y}^2) + V,$$

   where the "effective potential" V is energy dependent:

$$V = \frac{1}{2} m \frac{\rho^2 \dot{\rho}^2}{R^2 - \rho^2} = \frac{E \rho^2}{R^2}.$$

   Therefore, the motion *also* appears as a two-dimensional harmonic motion whose frequency Ω depends on the *total energy E*,

$$\Omega^2 = \frac{2E}{mR^2}.$$

4. Of course, if the square of the velocity is a constant in the *curved* four-dimensional space, this is not the case if one visualizes the phenomenon in a Euclidean plane, as above.
5. The simplicity of the result is intuitive. Quite obviously, as one can see in the definition (7.3), the symmetry of the problem is much larger than the sole rotation in $\mathcal{R}^3$. There is a rotation invariance in $\mathcal{R}^4$. The solutions of maximal symmetry correspond to a uniform motion on a circle of radius R in a plane whose orientation is arbitrary in $\mathcal{R}^4$. The whole set of solutions is obtained by projecting these particular solutions on planes of $\mathcal{R}^3$, which leads to the elliptic trajectories that we have found.

Conjugate Momenta and the Hamiltonian

The calculation of the conjugate momenta and the Hamiltonian is straightforward in this example. We obtain

$$p_\rho = m\dot{\rho}\,\frac{R^2}{R^2 - \rho^2}, \quad p_\theta = m\rho^2\dot{\theta}, \quad p_\phi = m\rho^2\sin^2\theta\,\dot{\phi}, \tag{7.148}$$

and the Hamiltonian

$$H = \frac{1}{2m}\left(p_\rho^2\,\frac{R^2 - \rho^2}{R^2} + \frac{p_\theta^2}{\rho^2} + \frac{p_\phi^2}{\rho^2\sin^2\theta}\right). \tag{7.149}$$

Chapter 8
Gravitational Waves

Even a blind man could see what the god meant.
Aristophanes, *Plutus (388 BC.)*

8.1 Evolution of Space-Time

In the previous chapter, we calculated three unexpected effects at the beginning of 20th century: the precession of Mercury's perihelion, the deviation of light rays by matter, and the dilation of proper time, which quickly became the celebrities of general relativity. Some are now part of our daily lives. Knowing that the motion of a free body in a curved space satisfies the principle of equivalence, we were able, thanks to the approximation (7.70) of the Schwarzschild metric, to perform correct calculations because gravitation is an extremely weak interaction. To continue, we need to take one more step in the theory.

General relativity aims to explain the gravitation and structure of the universe by imposing the principle of equivalence through a curved space-time with four dimensions. The curvature of this space-time comes from the various massive objects that are present. Now the gravitational structure of space-time arises from its curvature. The curvature therefore determines its own structure in space-time. This is a major mathematical difficulty of this theory which is by nature *non-linear*.

Our intention in this text is not to explain this theory. Nevertheless, we can sketch out the main points from what we have seen above and describe what is within our reach here.

Einstein's equations, which are its formulation, link the energy-momentum distribution $T_{\mu\nu}$ (or Tensor) to the curvature of space-time $G_{\mu\nu}$:

© The Author(s), under exclusive license to Springer Nature Switzerland AG 2023
J.-L. Basdevant, *Variational Principles in Physics*,
https://doi.org/10.1007/978-3-031-21692-3_8

$$G_{\mu\nu} \equiv R_{\mu\nu} - g_{\mu\nu}R = \frac{8\pi G}{c^4}T_{\mu\nu}, \tag{8.1}$$

G is the gravitational constant and c is the velocity of light in the vacuum, $g_{\mu\nu}$ is the metric tensor (here we adopt the $(-, +, +, +)$ signature).

The $T_{\mu\nu}$ energy-momentum tensor (also called stress-energy tensor) generalizes a simple point mass distribution. Its components consist of:

$$\begin{pmatrix} T_{00} & T_{0j} \\ T_{i0} & T_{ij} \end{pmatrix} = \begin{pmatrix} \text{energy density}/c^2 & \text{flux of energy density} \\ \text{momentum density} & \text{flux of momentum} \end{pmatrix}.$$

The Einstein tensor $G_{\mu\nu}$ is obtained as follows.

- We first work with the Riemann tensor $R^{\rho}_{\lambda\mu\nu}$ which will describe the curvature of space-time and which is directly connected to the metric tensor $g_{\mu\nu}$ using the Christoffel symbol $\Gamma^{\rho}_{\mu\nu} = (1/2)g^{\rho\lambda}\left(\partial_{\nu}g_{\mu\lambda} + \partial_{\mu}g_{\nu\lambda} - \partial_{\lambda}g_{\mu\nu}\right)$ that we saw in (7.16) in the previous chapter, and its derivatives, by

$$R^{\rho}_{\lambda\mu\nu} = \partial_{\lambda}\Gamma^{\rho}_{\mu\nu} - \partial_{\mu}\Gamma^{\rho}_{\lambda\nu} + \Gamma^{\sigma}_{\mu\nu}\Gamma^{\rho}_{\sigma\lambda} - \Gamma^{\sigma}_{\lambda\nu}\Gamma^{\rho}_{\sigma\mu} \tag{8.2}$$

- The Ricci Tensor $R_{\mu\nu}$ is obtained by contracting two indices of the Riemann Tensor: $R_{\mu\nu} \equiv R^{e}_{e\mu\nu}$. R is the corresponding curvature scalar: $R = g^{\mu\nu}R_{\mu\nu}$.

Variational Formulation

Einstein's equations can be obtained with a variational principle and the *Einstein-Hilbert action* S_{EH} which describes the dynamics of the gravitational field:

$$S_{EH} = \frac{c^4}{16\pi G} \int d^4x \, R\sqrt{(|g|)}, \tag{8.3}$$

where R is the scalar curvature of Ricci and g is the determinant of the metric tensor.

The constant $(c^4/16\pi G)$ is chosen so that one can simply add to it the actions that describe the matter to get the Einstein equations.

This (important) discovery is due quasi-simultaneously to Hilbert and Einstein.[1]

This topic is detailed by Sean Carroll.[2] Simple presentations can be found in references [17] (annex B), [18] page 438, and in Feynman's book.[3]

---

[1] One can find the details and passions that it revealed in the article by Dieter W. Ebner, arXiv:physics/0610154v1.

[2] Carroll, Sean M. (2004), *Spacetime and Geometry: An Introduction to General Relativity*, San Francisco: Addison-Wesley.

[3] Richard P. Feynman, *Feynman Lectures on Gravitation*, Addison-Wesley, (1995).

Einstein's equations are therefore a set of ten nonlinear partial differential equations, which are very difficult to solve completely. We'll benefit here of the fact that gravitation is an extremely weak interaction, and in many cases it can be treated in a perturbative way. Of course this is the dominant interaction of the vast majority of astrophysical phenomena, but most manifest themselves in such a way that their local description by Newtonian mechanics, in Minkowski space, is excellent. Note that electromagnetism and the physics of electromagnetic radiation, of indisputable importance in the cosmos, are in conformity with general relativity because this theory is, by construction, Lorentz invariant and that the mass of the photon is zero (see [17] Sect. 2.2). Let us point out that the calculations we have made previously come on the one hand from what Schwarzschild found in 1916, a particular solution of the Einstein equations in the vicinity of a large mass and, on the other hand, that at great distance from such a mass, the objects that populate the universe follow Newton's law in a Minkovski space.

Indeed, the circumstances in which the generality of Einstein's equations is unavoidable are reduced to the presence of ultra-dense clusters. One can obviously think of the origin of the Big Bang, but as soon as the recombination, 380,000 years later, the Newtonian approximation becomes quite suitable. The real objects for which one cannot escape a complete and difficult theory are the black holes themselves in their own structure.

It is useful to analyze the Schwartzschild formula (7.66) and how it was useful to us. As we have said, the correction of the curvature of space-time at the perihelion of Mercury can be calculated perturbatively because the effect is small. We could not have done this type of calculation at distances in the order of the Schwarzschild radius of the source. This metric has a singularity within the radius of Schwartzschild $r = r_S$. We obviously cannot calculate the deviation of light rays, nor any electromagnetic (or gravitational) phenomena in the $r \leq r_S$ region, the physics inside a black hole or in its vicinity, as in the Fig. 8.1,[4] is entirely in the realm of general relativity.

It remains that the question of the evolution of the curvature of space-time is inevitable. How does it occur, and especially is it detectable in another way than by observing the motion of the stars? How does it spread from one star to another? Let us recall that Newton himself was troubled by the apparent instantaneity of the action of the sun's gravity on Earth. Halley's results on the comet that bears his name seemed to confirm that this gravitational field was acting instantaneously at all times. Poincaré had already noticed in 1905 that special relativity suggested the existence of "gravific waves"[5] which would propagate.

---

[4] Kazunori Akiyama, Antxon Alberdi et al. *First M87 Event Horizon Telescope Results I, The shadow of the supermassive Black Hole*, The Astrophysical Journal, vol.875, 1, 2019, p. L1; *First M87 Event Horizon Telescope results VIII: magnetic field structure near the event horizon.* Astrophysical Journal Letters. Published online March 24, 2021.

[5] H. Poincaré, *On the dynamics of the electron*, Proceedings of the Académie des Sciences, t. 140, p. 1507 (5 June 1905).

Fig. 8.1 Left: First image of a black hole, published by the Event Horizon Telescope collaboration in 2019. This is the supermassive black hole M87* at the heart of the galaxy M87. The image shows the shadow of the horizon of the black hole events M87* in contrast with its accretion disk and what is called its photon sphere. Right: Recent image, Event Horizon data from March 2021 reveals twisted magnetic fields around the black hole M87*. Credit ESO, Event Horizon Telescope Collaboration

Within the framework of general relativity, Einstein was led to predict in 1916 the existence of gravitational waves, produced by these evolutions of the curvature of space-time. These waves were to be emitted in the evolution of the curvature, in a manner similar to the emission of electromagnetic waves that results from the motion of electrically charged objects. The subject remained for a long time a source of speculation (even from Einstein at the beginning) until the spectacular response given by the detection of these waves which interest us here.

8.2 Detection of Gravitational Waves

The first gravitational waves were observed experimentally a century after Einstein's prediction, during a direct detection of these waves on September 14th, 2015, confirmed on February 11th, 2016, by the detectors and physicists of the LIGO laboratory.[6] The emission, named GW150914,[7] came from the rapid spiral rotation of two black holes, located at 400 Mpc, or 1.3 billion light-years, of respective masses

---

[6] At the time of the occurrence, the Virgo gravitational wave detector was at stop for an improvement of its equipment, but all its researchers have worked on the data. The LIGO detector was on standby and was not supposed to resume its observations until September 18th, only four officials were aware of the reception of signals at the time it occurred. These four managers obeyed the working charter of the collaboration (more than 1000 physicists) and they only announced it to the rest of the collaboration four days later for all the data processing work. Full details, illustrated, of the collaboration between LIGO and Virgo Laboratories can be found at http://www.ligo.org and (http://www.virgo.infn.it).

[7] Abbreviation of "Gravitational Wave", September 14th, 2015; Abbott et al., *Observation of Gravitational Waves from a Binary Black Hole Merger*, Physical Review Letters, American Physical Society, vol. 116, no 061102, February 2016. This 12-page, very unusual length Physical Review Letters article contains 5 pages of names of physicists as well as institutions of the collaboration. The article itself contains absolutely all the information about the theoretical design, the experimental realization, and the results of this detection, with, of course, all the useful references.

$M = 36 M_\odot$ and $M = 29 M_\odot$ and their 'coalescence" or merged into a single black hole of $M = 62 M_\odot$. Considerable energy of 3, $M_\odot c^2$ was released in this fusion, binding energy released by the gravitational fusion of the binary system and carried away by the gravitational wave.[8] As pointed out by Thibault Damour, it was also the first direct evidence of black holes and the proof of the dynamics of space-time, by the fusion of two black holes, which is a confirmation of Einstein's theory[9] This is certainly the most significant basic research effort, both theoretical and experimental, in history. These observations were crowned with the award of the Nobel Prize in Physics in 2017 to Rainer Weiss, Barry C. Barish and Kip Thorne, "for their decisive contributions to the LIGO detector and to the observation of gravitational waves."

The detection of gravitational waves by a collaboration between LIGO—which has two sites located 3,000 km apart, one in Hanford, Washington, the other in Livingston, Louisiana—and Virgo which has a site near Pisa in Italy, will remain as one of the most important and significant physical discoveries of the beginning of the 21st century, together with that of the Higgs Boson, discovered in 2012 at the CERN LHC, expected fundamental particle of the standard model of fundamental interactions, which was predicted by Peter Higgs in 1964.

The signals received from GW150914 are shown in the Fig. 8.2. They come from the two LIGO observatories, Livingston and Hanford, which are 3,000 km apart, that is to say at a time interval of 7 ms one after the other.

They were followed and confirmed by a series of similar observations from the LIGO-Virgo collaboration,[10] from the strong gravitational interaction between heavy stars, pairs of black holes or neutron black hole-star systems. It is very informative to read the article on the discovery of GW150914 mentioned above, it contains considerable information on all aspects of this discovery, including the design of the detectors, and the measurement results. The signal was seen in Livingston 7 ms before it reached Hanford, which helps direct its provenance. We must emphasize the prodigious inventiveness and creativity of the physicists who designed the LIGO and Virgo detectors, both theoretically and experimentally.

In addition to the discovery itself, this observation opens up a field of observation in the universe infinitely farther than what we know now, since our universe is extremely transparent to gravitational waves because of their very weak interaction with matter.

---

[8] This amount of energy, which comes from a subtraction done by a child between the masses expressed in solar masses, is totally unimaginable in the solar system.

[9] "The most important is the proof of the existence of black holes", on lemonde.fr, February 11, 2016 (https://www.lemonde.fr/cosmos/article/2016/02/11).

[10] Full details of the results from the LIGO and Virgo laboratories can be found at (https://www.ligo.org/news.php) and (https://www.virgo-gw.eu).

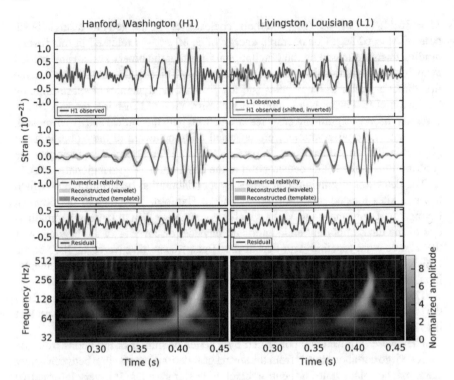

Fig. 8.2 The GW150914 gravitational event observed by the Hanford (H1 left panel) and Livingston (L1, right panel) LIGO detectors. The times shown horizontally correspond to September 14, 2015, at 09:50:45 UTC. For easy reading, all time series are filtered with a 35 to 350 Hz band-pass filter to suppress large fluctuations outside the most sensitive frequency band of the detectors. Vertically, the amplitude h of the strain in units 10^{-21}. *Top row*: left amplitude in H1, right amplitude in L1. GW150914 arrived first in L1 and 6.9 ms later in H1; For ease of visual comparison, H1 data is also shown, right with L1 data, offset in time by this amount and with an opposite sign (to take into account the relative orientation of the detectors). *Middle Row*: Gravitational strain projected on each detector in the 25–350 Hz band. The lines come from an optimized relativistic numerical calculation of the waveform with parameters compatible with those obtained by GW150914, confirmed at 99.9% by an independent calculation. Shaded areas are credible at 90% for two independent reconstructions of the waveform. One (in dark grey) models the signal using binary waveforms of black hole models. The other (light grey) does not use an astrophysical model, but rather calculates the deformation signal as a linear combination of Gaussian sinusoidal wavelets. These reconstructions have an overlap of 90%. These data treatments can be found in B.P. Abbott et al., Phys. Rev. Lett. 116, 241102 (2016). *Bottom row*: Time dependence of the frequency, and the intensity, of the stress signal shown above, we see a growth in time before coalescence, qualitatively in accordance with what we calculate below. As the coalescence approaches, the gravitational signal has a characteristic behaviour: both its frequency and its amplitude increase. (On the left Hanford, on the right, Livingston.) (Photo credit B.P. Abbott et al., Phys. Rev. Lett. 116, 241102 (2016))

8.3 Gravitational Waves

We do not wish to carry out here a complete theoretical study of gravitational waves. We will refer to specialized literature, but many of the key qualitative and quantitative aspects are easy to understand. We are interested in the nature of these waves, the orders of magnitude involved, the mechanism of their emission and their interaction with matter.

In the following, we will explain the nature and origin of the *amplitude* of the waves in these data, including its cause, its value, its evolution and its observation during the interaction of gravitational waves with the devices of the detectors.

This will allow us to understand the physics of the double pulsar, discovered in 1974 and analyzed by Taylor and Hulse who were awarded the Nobel Prize in 1993.

8.3.1 Quadrupole Waves; Orders of Magnitude

If the emission of gravitational waves is, in principle, a complicated problem, in many cases it can be reached in the Newtonian approximation, because we are only interested in the intensity, the structure and the propagation of these waves far from their source, in areas where the geometry is weakly deformed compared to Minkowski's. These waves, as we shall see, propagate at the speed of light. They are waves of space-time, They interact very weakly with matter. Their emission can be detected, but their passage through the cosmos is practically free, apart from the phenomena we have seen for light, such as the curvature of the light rays.

The order of magnitude of the brightness of the gravitational wave sources, the power they emit, is interesting. In the most "naive" case at present, of the emission by a rotating *binary system* that is, two masses orbiting around one another, the order of magnitude from the quantities involved is

$$\mathcal{L} \sim \frac{G}{c^5} s^2 \omega^6 M^2 R^4$$

where M and R are the mass and radius of the system, ω its pulsation, s its asymmetry factor (a spherical mass has a zero asymmetry $s = 0$, a dumbbell system an asymmetry $s = 1$), G is the Newton constant and c the speed of light.

With everyday objects, this effect is undetectable. But if we reformulate this expression using the orders of magnitude of systems that, *actually*, emit detectable waves, that is to say by returning to more significant quantities for this type of system, that is the speed $v = \omega R$, the Schwarzshild radius $R_S = 2GM/c2$, the same expression is rewritten as

$$\mathcal{L} \sim \frac{c^5}{G} s^2 \left(\frac{v}{c}\right)^6 \Xi^2 \tag{8.4}$$

where $\Xi = R_S/R$ is called the system's "relativistic compactness parameter".

This parameter is close to $\varXi = 1$ for black holes, to $\varXi \simeq 0, 2$ for neutron stars, to $\varXi \simeq 10^{-3}$ for white dwarfs, to $\varXi \simeq 10^{-6}$ for stars like the sun, and it goes down to $\varXi \simeq 10^{-10}$ for a planet like the Earth. We see that the potential emitters of observable gravitational waves are the compact objects.

The first factor of (8.4), is essential It is built with two of the three fundamental constants, G and c, it is the "Planck luminosity" defined by

$$\mathcal{L}_P = \frac{c^5}{G} \approx 3, 6 \ 10^{52} \text{ watt.} \tag{8.5}$$

This considerable maximum value, in the order of 10^{26} times that of the Sun, is *effectively reached* by black holes. We will show it in Sect. 3.7 It takes place only during the brief time of the fusion of two black holes, a fraction of a second. It's one of those mythical numbers of general relativity, like Planck's strength $F_P = (c^4/4G)$, coefficient of passage from the Einstein tensor to energy-momentum tensor in Eq. (8.1). But it is now accessible to our intuitive perception.

What is perhaps more astonishing is to see how our intuition is troubled by the coming of the third fundamental constant, the Planck constant $\hbar$. The latter, present in all the usual effects of the other three fundamental interactions, such as matter at our scale, leads to a Planck length of ℓ_P and a Planck time of t_P of

$$\ell_P = \sqrt{\frac{\hbar G}{c^3}} \sim 1.6, 10^{-35} \text{m} \,, \quad t_P = \sqrt{\frac{\hbar G}{c^5}} \sim 5.4, 10^{-44} \text{s; ,}$$

totally out of our intuitive apprehension.

8.3.2 Formation and Propagation of the Gravitational Waves

In general, gravitational waves fill space-time. We are going to focus here on a relatively simple process, but a very important one since it concerns the formation and then the detection of the waves produced and observed so far. This is the motion and evolution of a binary system, two objects—two black holes typically- rotating under the influence of their mutual gravitational attraction, which radiate in this motion and lose energy until they merge on one another and produce a more massive black hole.

Two facts are important.

- The first is that this simple physical process to describe and understand is the same one that Einstein had imagined. The formula of the quadrupole obtained by him in 1916 links the amplitude h of the wave emitted by a physical system to the variation of its quadrupole moment Q (recall that the gravitational field, only attractive, has no dipole moment):

$$h = \frac{2G}{c^4 r} \ddot{Q}(t - r/c), \tag{8.6}$$

where the quadrupole moment Q of a mass distribution $\rho(\mathbf{r})$ is

$$Q_{ij} = \int \rho(3r_i r_j - \|\mathbf{r}\|^2 \delta_{ij}) \, d^3\mathbf{r}. \tag{8.7}$$

We will see that the parameter h of (8.6) determines the gravitational deformation (tidal effect) that the passage of the wave causes on the distance between two points (in free fall) of the receiver. It is this parameter of *deformation* which is measured and shown in Fig. 8.2. We will show in Sect. 3.5 how the theory predicts that.

The smallness of the factor $2G/c^4 \approx 1.65 \, 10^{-44} \mathrm{m}^{-1} \mathrm{kg}^{-1} \mathrm{s}^2$, cause of the smallness of h, illustrates what is called the "stiffness" of space-time. It must be compensated by large variations of the quadrupole moment in order to produce detectable effects (any attempt on a human scale, such as gigantic wagons on a merry-go-round, would give tiny results).

- On the other hand, concerning the source of the wave, we will examine below a binary system of massive stars, typically black holes or pulsars in relative motion, in the Newtonian approximation which gives an extremely precise result. We will see at the end of this chapter the amazing observation of Taylor and Hulse that led them to the discovery of the double Pulsar PSR B1913+16 where they measured the loss of rotational energy of two orbiting pulsars with an accuracy that is consistent with the Newtonian calculation of motion, and the prediction of general relativity for the loss of energy by the corresponding gravitational radiation.

8.3.3 Linearization of Einstein's Equations

Consider a region of space-time where the geometry is little distorted compared to that of Minkowski by the passage of waves of low amplitude. Einstein's equations can be linearized by developing in the first order the effect of the disturbance. In first order the metric has the form

$$g_{\mu\nu} = \eta_{\mu\nu} + h_{\mu\nu}, \quad \text{with } |h_{\mu\nu}| \ll 1. \tag{8.8}$$

where $\eta_{\mu\nu}$ is the Minkowski metric and $h_{\mu\nu}$ is dimensionless. (We follow the method used in [17], Chap. 7.)

The linearized Christoffel symbols are then

$$^{(1)}\Gamma^\lambda_{\mu\nu} = \frac{1}{2}\eta^{\lambda\sigma}(\partial_\mu h_{\sigma\mu} + \partial_\nu h_{\mu\sigma} - \partial_\sigma h_{\mu\nu}) \tag{8.9}$$

$$^{(1)}R^{\rho}_{\lambda\mu\nu} = \partial_\lambda {}^{(1)}\Gamma^{\rho}_{\mu\nu} - \partial_\mu {}^{(1)}\Gamma^{\rho}_{\lambda\nu}. \tag{8.10}$$

We deduce the Ricci tensor at first order

$$^{(1)}R_{\mu\nu} = \partial^\sigma \partial_\mu h_{\nu\sigma} - \frac{1}{2}\partial^\sigma \partial_\sigma h_{\mu\nu} - \frac{1}{2}\eta^{\rho\sigma}\partial_\mu \partial_\nu h_{\rho\sigma}. \tag{8.11}$$

Defining

$$\bar{h}_{\mu\nu} = h_{\mu\nu} - \frac{1}{2}h\eta_{\mu\nu} \quad \text{with} \quad h \equiv \eta^{\rho\sigma}h_{\rho\sigma}, \tag{8.12}$$

Einstein' equation at that order becomes

$$\partial^\sigma \partial_\sigma \bar{h}_{\mu\nu} - \eta_{\mu\nu}\partial_\rho \partial_\sigma \bar{h}^{\rho\sigma} - \partial_\nu \partial_\rho \bar{h}^{\rho}_{\nu} - \partial_\mu \partial_\rho \bar{h}^{\rho}_{\nu} = -2\frac{8\pi G}{c^4}T_{\mu\nu}. \tag{8.13}$$

Choice of Gauge

In an infinitesimal coordinate transformation $x'^{\mu} = x^{\mu} + \xi^{\mu}(x)$, $h_{\mu\nu}$ becomes

$$h'_{\mu\nu} = h_{\mu\nu} - \partial_\mu \xi_\nu - \partial_\nu \xi_\mu.$$

Thus, in a similar way to electromagnetism, we can choose a system of coordinates that ensures the condition

$$\partial_\mu \bar{h}^{\mu\nu} = 0. \tag{8.14}$$

This is called the Lorenz Gauge, by analogy with electromagnetism.
The equations of the fields then become very simple[11]

$$\partial^\sigma \partial_\sigma \bar{h}_{\mu\nu} = -2\frac{8\pi G}{c^4}T_{\mu\nu}. \tag{8.15}$$

In the absence of matter, the energy-momentum tensor is zero. The waves propagate at the speed of light in a flat Minkowski space. An important difference with electromagnetic waves, which propagate fields, is that gravitational waves are associated with the metric tensor, which is twice covariant whereas the electromagnetic four-vector is once covariant.

8.3.4 Polarization: Transverse Traceless Waves

The amplitude must meet the above gauge conditions. Consider a wave propagating in the direction of Oz. Among the ten components of the symmetric tensor $h_{\mu\nu}$ a

---

[11] For all practical details, see [17–19].

suitable choice of the four functions $\xi^\mu(x)$ above, cancels 4 combinations of the $h_{\mu\nu}$ in the vacuum:

$$h = \eta^{\mu\nu} h_{\mu\nu} = 0, \quad h_{0i} = 0 \qquad (8.16)$$

This defines what is called the TT gauge (Transverse and Traceless) Another similar set of considerations (see [17] chap.7 §3) shows that we are reduced to four non-zero components. In this case (propagation along Oz), these components are H_{xx}, $H_{yy} = -H_{xx}$, and $H_{xy} = -H_{yx}$.

The wave is transverse and characterized by two polarizations.

The mode H_{xx} is called polarization $+$ and the mode H_{xy} is called polarization $\times$.

8.3.5 Detection of the Waves

Consider a free test particle initially at rest. In the TT gauge, this particle will travel on a geodesic determined by the metric

$$\mathbf{g} = -c^2 dt^2 + (\delta_{ij} + h_{ij}^{TT}) dx^i dx^j. \qquad (8.17)$$

The measurement of the amplitude of the gravitational wave is done by considering two free particles A and B and by determining their distance L from the half of the proper time of one of them, A, the round-trip time that elapses between the emission of a light ray to the other, B, and the return from B to A

$$L = \frac{1}{2} c (\tau_A^{recep} - \tau_A^{emis}). \qquad (8.18)$$

For two particles at rest, we obtain, in the TT gauge,

$$L = \sqrt{g_{ij} \Delta x^i \Delta x^j} \quad \text{where} \quad \Delta x^i = x_B^i - x_A^i. \qquad (8.19)$$

The relative variation in distance between A and B at the passage of the gravitational wave is therefore

$$\frac{\delta L}{L_0} = \frac{1}{2} h_{ij}^{TT} n^i n^j \qquad (8.20)$$

where n^i is the unit vector ($\delta_{ij} n^i n^j = 1$) such that $\Delta x^i = L_0 n^i$ before the wave passes.

To visualize the effect of the gravitational wave, it is convenient to insert the coordinates

$$X^i = x^i + \frac{1}{2} h_{ij}^{TT}(t, 0) x^j$$

that will fluctuate as the wave passes. We see on this expression how the amplitude h intervenes in the order of magnitude of the effect to be measured. In current detections, this order of magnitude is of the order of $h \sim 10^{-21}$ (Fig. 8.2), that we will calculate on a model. This deformation effect, tidal effect, is of the same order of magnitude as the size of an atom compared to the Earth-Sun distance.

Consider a monochromatic wave propagating along Oz. The wave corresponding to the $+$ polarization is of the form

$$h_{xx}^{TT} = h_+ \sin(\omega t - kz), \quad h_{yy}^{TT} = -h_+ \sin(\omega t - kz), \quad t \geq 0. \tag{8.21}$$

A set of test particles, all in free fall, located on a circle of radius a in the plane $z = 0$ undergoes a variable distortion in time according to

$$X = a[\cos\theta + \frac{1}{2}h_+ \sin(\omega t)\cos\theta], \quad Y = a[\sin\theta + \frac{1}{2}h_+ \sin(\omega t)\sin\theta] \tag{8.22}$$

For the polarization $\times$, the gravitational wave is of the form

$$h_{xy}^{TT} = h_{yx}^{TT} = h_\times \sin(\omega t - kz), \tag{8.23}$$

and the deformation

$$X = a[\cos\theta + \frac{1}{2}h_\times \sin(\omega t)\sin\theta], \quad Y = a[\sin\theta + \frac{1}{2}h_\times \sin(\omega t)\cos\theta]. \tag{8.24}$$

These deformations are represented (very exaggerated: with $h \sim 0.5$) in Fig. 8.3.

The two polarizations correspond to distinct deformations, and the interferences between the two modes are amplified.

For this reason, the LIGO and Virgo experiments, which must reach a sensitivity of $h \sim 10^{-21}$, use Michelson interferometers with very high-performance lasers, multi-kilometre arms (Fig. 8.4) and Fabry-Pérot devices that multiply the back and forth distances of light in each arm of the formula (8.19).

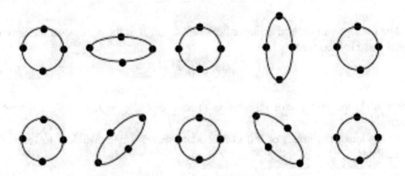

Fig. 8.3 Deformation, at successive moments, of a circle of test free particles at the passage of a gravitational wave. Top: polarization $+$, bottom: polarization $\times$

Fig. 8.4 Observatories of LIGO, at Hanford (left): the arms are 4 km long, and of Virgo at Pisa (right): the arms are 3 km long

In the Virgo interferometer, shown below, the purified laser beam, first split in two by a separating blade, is reflected on mirrors in each of the two perpendicular arms, one to the north (NE), the other to the west (WE). These two 40 kg mirrors are perfectly reflective and allow 99.999% of the incident light to pass through. The two reflected beams recombine on the mirror at 45°, producing an interference pattern. Each beam travels back and forth in reflecting on mirrors. It eventually comes out and crosses the other beam with which it recomposes itself. If both beams travelled the same distance, the waves light is perfectly compensated and the output of the interferometer remains obscure. However, if a gravitational wave passes, it shortens one arm and lengthens the other. The two beams recompose with a slight phase shift and light comes out of the interferometer. This is observed by a photodetector, which converts the incident light into an electric current, which is then amplified and digitally recorded for analysis (Fig. 8.5).

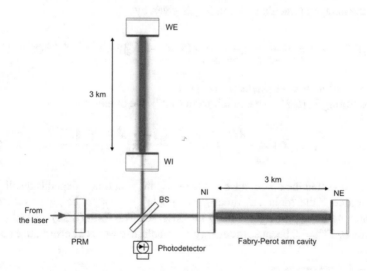

Fig. 8.5 Simplified optical diagram of the Virgo interferometer. One tube is directed to the north: NE, and the other to the west: WE. The widening of the lines represents the amplification of the laser rays by the Fabry-Pérot cavities. (Credits: Virgo Collaboration)

This detection is in itself a conceptual and experimental feat as one can see in the article *Virgo interferometer*[12]

8.3.6   Generation of Gravitational Waves

After propagation and detection, let's see how gravitational waves are produced by the presence of matter.

General Aspect of the Problem

We seek to solve the linearized Einstein equations linearized with the energy-momentum tensor:

$$\partial^\sigma \partial_\sigma \bar{h}_{\mu\nu} = -\frac{16\pi G}{c^4} T_{\mu\nu} \tag{8.25}$$

which is analogous to the electromagnetic equation (6.40) in the presence of a source.

In the absence of incoming gravitational waves, the (8.25) solution is formally written as

$$\bar{h}_{\mu\nu} = -\frac{16\pi G}{c^4} \int d^4y\, \mathcal{G}_{ret}(x-y) T_{\mu\nu}(y) \tag{8.26}$$

where the delayed Green's function $\mathcal{G}_{ret}$ is given by

$$\mathcal{G}_{ret}(x^\lambda - y^\lambda) = -\frac{1}{4\pi \parallel \vec{x} - \vec{y} \parallel} \delta[\parallel \vec{x} - \vec{y} \parallel -(x^0 - y^0)]\Theta(x^0 - y^0), \tag{8.27}$$

and is a solution of $\partial^\sigma \partial_\sigma \mathcal{G}_{ret}(x^\lambda) = \delta^4(x^\lambda)$.

Substituting in (8.26) the explicit form (8.27), we obtain

$$\bar{h}_{\mu\nu}(t, \vec{x}) = -\frac{4G}{c^4} \int d^3y\, \frac{T_{\mu\nu}(ct- \parallel \vec{x} - \vec{y} \parallel, \vec{y})}{\parallel \vec{x} - \vec{y} \parallel}. \tag{8.28}$$

This shows that the disturbance of the metric in $\vec{x}$ at time t depends on all points of the source at the "delayed" times $t_r = t- \parallel \vec{x} - \vec{y} \parallel /c$, i.e., universe lines of the source with the past light cone of the observer. The form (8.28) obeys Lorenz's condition $\partial_\mu T^{\mu\nu} = 0$ because the energy-momentum tensor is a perturbative quantity

---

[12] http://public.virgo-gw.eu/the-virgo-collaboration, Reflections of physics, Issue 52, February 2017.

of the same order as the $h_{\mu\nu}$ and the corrective terms from the covariant derivative are of second order.

Asymptotic Solution Away from Source

As far as we are concerned, the observer, in $\overrightarrow{x}$, is located very far from the source. By setting the origin of the coordinates inside the source, we can replace the denominator $\| \overrightarrow{x} - \overrightarrow{y} \|$ by the distance of the observer $r = \| \overrightarrow{x} \|$. To approximate t_r by $t - r/c$ in the energy-momentum tensor, assumes a slowly varying source. The expression (8.28) becomes under these conditions

$$\bar{h}_{\mu\nu}(t, \overrightarrow{x}) = -\frac{4G}{c^4 r} \int d^3 y \, T_{\mu\nu}(ct - r, \overrightarrow{y}) \tag{8.29}$$

Let us now consider the spatial components of this expression. These are the only ones we need to work with in the TT gauge. Thanks to the energy-momentum tensor conservation, the integrals $\int d^3 y \, T_{ij}(y)$ can be expressed simply.

The integral of the components of T^{ij} are, after integrating by parts

$$\int d^3 y \, T^{ij} = \int d^3 y \, \partial_k(y^i T^{kj}) - \int d^3 y \, y^i \partial_k T^{kj},$$

where the first term gives a surface term that cancels out because the source is compact. Energy-momentum tensor conservation implies

$$\partial_k T^{kj} = -\partial_0 T^{0j}.$$

By substituting in the previous relation, we have

$$\int d^3 y \, T^{ij} = \frac{d}{dt} \int d^3 y \, y^i T^{0j} = \frac{1}{2} \int d^3 y (y^i T^{0j} + y^j T^{0i})$$

$$= \frac{1}{2} \frac{d}{dt} \int d^3 y [\partial_k(y^i y^j T^{0k}) - y^i y^j \partial_k T^{0k}],$$

and, using the gauge condition, which gives $\partial_k T^{0k} = \partial_0 T^{00}$, one obtains the equality

$$\int d^3 y \, T_{ij} = \frac{1}{2} \frac{d^2}{dt^2} \int d^3 y \, y^i y^j T^{00}. \tag{8.30}$$

We thus obtain Einstein's formula: gravitational waves emitted by a distribution of matter are written

$$\bar{h}_{ij}(t, \vec{x}) = \frac{2G}{c^4 r}\left(\frac{d^2}{dt^2}I_{ij}(t - r/c)\right) \quad \text{where} \quad I_{ij} = \int d^3y\, \rho\, y_i y_j, \qquad (8.31)$$

I^{ij} is the quadrupole moment of the source, introduced by Einstein in 1916, or

$$h_= \frac{2G}{c^4 r}\ddot{Q}(t - r/c). \qquad (8.32)$$

The quadrupole moment Q of a continuous system of volume mass density $\rho(x, y, z)$ is

$$Q_{ij} = \int \rho(x, y, z)\,(3r_i r_j - \| \vec{r} \|^2\, \delta_{ij})d^3r.$$

8.3.7   Radiated Power

Gravitational radiation causes a decrease of the energy of the source. The calculation of this energy loss was used in particular in the case of the variation of the distance, thus of the rotation frequency, between the two sources of a binary Pulsar system, and constituted the first, and very remarkable, verification of the predictions of general relativity: the discovery of the double Pulsar PSR B1913+16 by Taylor and Hulse, 1993 Nobel Prize winner. We will show these results below.

In electromagnetism, the flux of energy radiated in the form of electro-magnetic waves is obtained from the energy-momentum tensor which varies quadratically with A_μ.

Similarly, in this case, we must evaluate the energy carried by gravitational waves from the energy-momentum tensor which must be quadratic in the perturbation $h_{\mu\nu}$.

We must therefore develop Einstein's equations to second order of perturbations:

$$^{(1)}G_{\mu\nu} + ^{(2)}G_{\mu\nu} = 8\pi G T_{\mu\nu} \qquad (8.33)$$

By shifting the term $h_{\mu\nu}$ to the right (see [17] Chap. 7.6) one gets the expression

$$^{(1)}G_{\mu\nu} = \frac{8\pi G}{c4}(T_{\mu\nu} + \tau_{\mu\nu}), \quad \text{i.e. } \tau_{\mu\nu} = -\frac{c^4}{8\pi G}\,^{(2)}G_{\mu\nu} \qquad (8.34)$$

where the term $\tau_{\mu\nu}$ quadratic in $h_{\mu\nu}$ is "analogous" to an energy-momentum tensor, although it's not a real tensor. By averaging over a few wavelengths and integrating the flow of energy on a large sphere centered on the source and passing through the observer, we obtain a gravitational luminosity (or power) of the quadrupole

$$\mathcal{L}_G = \frac{G}{5c^5}\langle \dddot{Q}_{ij}\dddot{Q}^{ij}\rangle. \qquad (8.35)$$

where Q_{ij} is the traceless part of the quadrupole moment defined above.

Another approach comes from the analysis of Isaacson's effective energy-momentum tensor (See [19], Chap. 6, Sect. 5) and the h_+ and $h_\times$, h_{ij}^{TT} amplitudes. We get an energy flow of F in the direction of the observer

$$F = \frac{c^3}{16\pi G}\langle (h_+^2 + h_\times^2)\rangle \tag{8.36}$$

the average is also taken over a few wavelengths, so that $\langle h_+^2\rangle = \langle h_\times^2\rangle = (2\pi f)^2 h^2/2$ where f is the frequency of the waves

$$F = \frac{\pi c^3}{4G} f^2 h^2, \tag{8.37}$$

so that by integrating on a large sphere surrounding the source and using the quadrupole formula, we find the brightness (8.35).

We can now understand the change from the very large factor c^3/G of the expression (8.37) to the very small factor G/c^5 of (8.35). We discussed this above. To have a realistic order of magnitude of this gravitational power, or brightness, we must related it to the "natural" orders of magnitude of this effect in the bodies that emit it. We must compare the speed v with the speed of light c, the radius and the Schwartzchild radius $R_S = 2GM/c^2$, a simple rearrangement of terms leads to rewrite this formula of brightness with the appropriate parameters, as in (8.4):

$$\mathcal{L} \sim \frac{c^5}{G} s^2 \left(\frac{v}{c}\right)^6 \left(\frac{R_S}{R}\right)^2$$

so that for systems or objects for which $s \sim 1$, $R \sim R_S$ and $v/c \sim 1$ which is the case for systems of black holes or neutron stars in close orbit, we obtain luminosities of the order of $\mathcal{L} \sim 10^{52}$W ,i.e. 10^{26} times the light power of the sun or 10^4 times that of all the stars of the current accessible universe, for a relatively short time, a fraction of a second.

8.4 Binary System

The calculation of the Newtonian motion of the two stars in gravitational interaction is simple and consistent with the observation.

8.4.1 Motion of Two Stars

In the Newtonian approximation, the motion of two bodies of masses M_1 and M_2 is written (a bold character $\mathbf{x}$ denotes a three-vector and dots the time derivatives: $\dot{\mathbf{x}}$ and $\ddot{\mathbf{x}}$)

$$M_1\ddot{\mathbf{x}}_1 = -\frac{GM_1M_2}{\parallel \mathbf{x}_1 - \mathbf{x}_2 \parallel^3}(\mathbf{x}_1 - \mathbf{x}_2), \quad M_2\ddot{\mathbf{x}}_2 = -\frac{GM_1M_2}{\parallel \mathbf{x}_1 - \mathbf{x}_2 \parallel^3}(\mathbf{x}_2 - \mathbf{x}_1). \quad (8.38)$$

The separation vector between the two stars $\mathbf{d} = \mathbf{x}_1 - \mathbf{x}_2$ satisfies the equation of motion

$$\ddot{\mathbf{d}} = -\frac{GM_{tot}}{d^3}\mathbf{d} \quad (8.39)$$

where $M_{tot} \equiv M_1 + M_2$ is the total mass of the system. The total energy of the system can be written in two ways:

$$E = \frac{1}{2}M_1\dot{\mathbf{x}}_1^2 + \frac{1}{2}M_2\dot{\mathbf{x}}_2^2 - \frac{GM_1M_2}{d} = \frac{1}{2}M_{tot}\dot{\mathbf{x}}_B^2 + \frac{1}{2}\mu\dot{\mathbf{d}}^2 - \frac{GM_{tot}\mu}{d} \quad (8.40)$$

where $\mathbf{x}_B \equiv (M_1\mathbf{x}_1 + M_2\mathbf{x}_2)/M_{tot}$ is the system center of mass and $\mu = M_1M_2/M_{tot}$ the reduced mass. In the following, we place ourselves in an (inertial) reference system where the c.m. remains localized at the origin. The total energy is then reduced to that of the relative motion which is equivalent to the motion of a fictitious mass particle μ in the potential $V(d) = -GM_{tot}\mu/d$.

If we suppose a circular motion of angular velocity Ω, the equation of motion (8.39) involves $\Omega^2 d = GM_{tot}/d^2$, from which we deduce

$$\Omega = \frac{(GM_{tot})^{1/2}}{d^{3/2}}. \quad (8.41)$$

The energy of the system is then

$$E = \frac{1}{2}\mu\Omega^2 d^2 - \frac{G\mu M_{tot}}{d} = -\frac{G\mu M_{tot}}{2d}. \quad (8.42)$$

The trajectories of the two stars are then given by

$$\mathbf{x}_1 = \frac{M_2}{M_1 + M_2}\mathbf{d}, \quad \mathbf{x}_2 = \frac{M_1}{M_1 + M_2}\mathbf{d} \quad (8.43)$$

By choosing x and y coordinates in the binary system plane, the origin being placed at the c.m. of the system, the components of $\mathbf{d}$ are of the form

$$\mathbf{d} : \{d\cos\Omega t, \ d\sin\Omega t\} \quad (8.44)$$

or:

$$x_1 = \frac{M_2}{M_{tot}} d \cos \Omega t, \quad y_1 = \frac{M_2}{M_{tot}} d \sin \Omega t$$

$$x_2 = \frac{M_1}{M_{tot}} d \cos \Omega t, \quad y_2 = \frac{M_1}{M_{tot}} d \sin \Omega t$$

8.4.2 Transmitted Gravitational Waves

The two stars are treated as point objects. The quadrupole moment of the binary system is simply

$$I_{ij} = \int d^3x \, \rho \, x_i x_j = M_1 (x_1)_i (x_1)_j + M_2 (x_2)_i (x_2)_j \qquad (8.45)$$

whose components are

$$I_{xx} = M_1 x_1^2 + M_2 x_2^2 = \mu d^2 \cos^2 \Omega t = \frac{1}{2} \mu d^2 (1 + \cos 2\Omega t)$$

$$I_{yy} = M_1 y_1^2 + M_2 y_2^2 = \mu d^2 \sin^2 \Omega t = \frac{1}{2} \mu d^2 (1 - \cos 2\Omega t)$$

$$I_{xy} = M_1 x_1 y_1 + M_2 x_2 y^2 = \mu d^2 \cos \Omega t \sin \Omega t = \frac{1}{2} \mu d^2 \sin 2\Omega t$$

By substituting in (8.31) we obtain the gravitational wave emitted by this binary system:

$$\bar{h}_{ij}(t, \mathbf{x}) = \frac{4G^2 \mu M_{tot}}{rdc^4} \begin{pmatrix} -\cos(2\Omega t_e) & -\sin(2\Omega t_e) & 0 \\ -\sin(2\Omega t_e) & \cos(2\Omega t_e) & 0 \\ 0 & 0 & 0 \end{pmatrix}$$

with $t_e \equiv t - r/c$, where r is the distance between the source and the observer, and t_e the time of emission.

The order of magnitude $h \sim (GM)^2/c^4 rd \sim r_S^2/rd$ of the amplitude obtained for the gravitational waves emitted in GW150914, in the previous calculation by taking the two initial black holes of the same mass $M \sim 30 M_\odot$ or Schwartzschild radius $r_S \sim 90\,\mathrm{km}$, and orbiting at $d \sim 7r_S = 630\,\mathrm{km}$, with a detection at a distance of $400\,\mathrm{Mpc} \sim 1.3\,10^{22}\,\mathrm{km}$ leads to an acceptable value of this amplitude $h \sim 10^{-21}$.

8.4.3 Binary System Energy Loss

Back to the case of a double system and the formula of the quadrupole (8.35). The gravitational brightness of the binary system is

$$\mathcal{L}_G = \frac{32}{5} \frac{G^4 \mu^2 M_{tot}^3}{c^5 d^5} \tag{8.46}$$

where M is the total mass, μ is the reduced mass, d is the distance of the two objects. The angular speed of rotation is $\Omega = (GM/d^3)^{1/2}$, with a period of $P = 2\pi/\Omega$ (Kepler's third law).

The emission of gravitational waves corresponds to a decrease in the energy of the system:

$$\dot{E} = -\mathcal{L}_G \tag{8.47}$$

This causes a gradual bringing together of the two bodies, a decrease in the distance d and the period P :

$$\dot{P} = -\frac{3}{2}\frac{P}{E}\mathcal{L}_G = -\frac{96}{5}\frac{G^3 \mu M_{tot}^2 P}{d^4}. \tag{8.48}$$

or

$$\dot{P} = -\frac{3}{2}\frac{P}{E}\mathcal{L}_G = -\frac{96}{5}\frac{G^3 \mu M_{tot}^2 P}{d^4}. \tag{8.49}$$

Using Kepler's law $\Omega = (GM/d^3)^{1/2}$, we can deduce a differential equation for the period:

$$\dot{P} = -\frac{96}{5}(2\pi)^{8/3} G^{5/3} \mu M_{tot}^{2/3} P^{-5/3} \tag{8.50}$$

Evolution of the Distance and the Frequency

The distance d between the two stars changes, according to the above,

$$\dot{d} = -\frac{64}{5}\frac{G^3 \mu M_{tot}^2}{d^3}, \tag{8.51}$$

which is directly integrated.

$$d^4 = \frac{256}{5}(G^3 \mu M_{tot}^2)(t_c - t) \tag{8.52}$$

where t_c is the fusion or coalescence time (to be determined).

The change in the frequency of the waves $f = 2\Omega/2\pi = 2/P$ is deduced by

$$\frac{\dot{f}}{f} = -\frac{\dot{P}}{P} = \frac{96}{5} \frac{G^3 \mu M_{tot}^2}{d^4}, \tag{8.53}$$

or

$$\dot{f} = \frac{96}{5} (\pi)^{8/3} G^{5/3} \mu M_{tot}^{2/3} f^{11/3}. \tag{8.54}$$

The time evolution of the frequency and amplitude of gravitational waves at the approach of the moment of fusion or coalescence t_c are deduced from that:

$$f \propto (t_c - t)^{-3/8} \quad h \propto \frac{1}{d} \propto (t_c - t)^{-1/4}. \tag{8.55}$$

As coalescence approaches, the gravitational signal has a characteristic behaviour: both its frequency and its amplitude increase.

One can see that on the results of LIGO-Virgo, on the lower part of Fig. 8.2, where are represented the evolution of the frequency of the waves detected in GW150914, at Hanford and Livingston. We observe the increase of this frequency, until the moment of coalescence, when the waves cease to be emitted.

8.5   Double Pulsar Discovery PSR B1913+16

The result of Mercury's perihelion had been Einstein's greatest scientific emotion. Eddington's results in 1919 on the deviation of light rays had amazed the world and instantly made Einstein's international scientific reputation. Many other measurements on time dilation took place in the following decades, including the remarkable experiment of Pound and Rebka on the red displacement of spectral lines in 1959. However, it is a result of exceptional precision on an unexpected phenomenon that was the first to be admitted as a proof of General Relativity when the 1993 Nobel Prize was awarded to Joseph H. Taylor , and Russell A. Hulse "For the discovery of a new type of pulsar, which has opened up new possibilities for the study of gravitation". The discovery was twofold. On the one hand Taylor and Hulse had discovered the first example of a double pulsar, on the other hand, the rotation of this system emitted gravitational energy and its rotation period decreased over time with an accuracy identical to that of the best theoretical calculations of General Relativity.

The discovery of the double pulsar PSR B1913+16 is in itself amazing. In 1974, Hulse and Taylor recorded pulsar signals on Arecibo's 305-meter reflector (destroyed in December 2020). These radio signals, instead of being perfectly regular as standard ones, varied by advancing or delaying repetitively with a period of 7.75 hours. They understood that they came from the fact that the pulsar in question was in orbit (binary) with another star, identified as another pulsar whose possible emissions did not reach the solar system. The effect of the rotation of the two pulsars was reflected in the frequency of the observed radio signal of the first pulsar. This was an important astronomical discovery. Taylor and Hulse were first able to establish with great

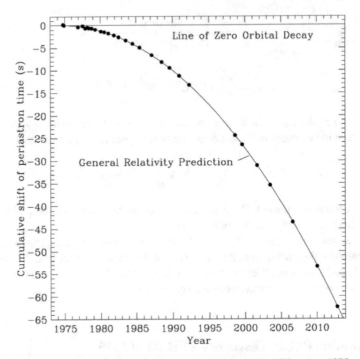

Fig. 8.6 Cumulative decrease in the period of orbit of PSR 1913B+16 between 1975 and 2014. The points are experimental measurements, the curve is the calculated value in general relativity. (T. Damour et N. Deruelle, Ann. Inst. Henri Poincaré, Vol. 44, No. 3, 1986). (Photo credit: The Astrophysical Journal, Volume 829, Number 1 (2016))

precision the parameters of the binary system, pulsar of mass $M_1 = 1.4414\, M_\odot$, satellite of mass $M_2 = 1.3857\, M_\odot$, of very eccentric orbit, periastre ~ 1.1, $R_\odot$, apoastre ~ 4.8, $R_\odot$, eccentricity $e = 0.617$. Since its discovery, the orbit has evolved in accordance with the predictions of general relativity: precession of the periastre of 4.22° per year, and decay of the semilong axis (of $1.95\ 10^6$ km) of 3.5 m per year.

The most interesting measure of precision is that of the 27, 907 s period, which decreases by 76 μs per year. This decrease comes from the system's loss of energy by gravitational wave emission. There are 1130 orbits described per year and the cumulative decrease of the period observed between 1975 and 2003 is shown in the Fig. 8.6 accompanied by the forecast calculated by general relativity.[13]

The relativistic calculation has been done and refined by several authors.[14] Apart from details too complex for this chapter, we find the structure of the expression

---

[13] See in particular J. M. Weisberg and Y. Huang, *Relativistic measurements from timing the binary pulsar PSR B1913+16*, The Astrophysical Journal, Volume 829, Number 1 (2016), and references.

[14] See Nathalie Deruelle, Jean-Pierre Lasota: Gravitational Waves, Odile Jacob "Sciences", March 2018; Damour T. and Deruelle N. 1986 AnIHP 44 263; Damour T. and Taylor J. H. 1991 ApJ 366 501; Damour T. and Taylor J. H. 1992 PhRevD 45 1840; Weisberg J. M. and Taylor J. H. 1981 GReGr 13 1; Weisberg J. M. and Taylor J. H. 2002 ApJ 576 942.

(8.55) taking completed into account the e eccentricity of the system, that is

$$\dot{P}^{GR} = -\frac{192\pi G^{5/3}}{5c^2}(\frac{P_b}{2\pi})^{-5/3}m_1 m_2(m_1 + m_2)^{-1/3}\left(1 + \frac{73}{24}e^2 + \frac{37}{96}e^4\right)(1 - e^2)^{-7/2}$$

(8.56)

where P_b is the observed period. By inserting the measured values for the parameters, this equation gives

$$\dot{P}^{GR} = -2,402\ 10^{-12} \quad \text{for} \quad \dot{P}^{exp} = -2,398\ 10^{-12}$$

(8.57)

which, by adding experimental errors, amounts to $\dot{P}^{exp}/\dot{P}^{GR} = 0,9983 \pm 0,0016$.

It will obviously be noted that the corrections of general relativity to the Newtonian formula (8.55) give a very remarkable precision of the result.

We can look at these values with other orders of magnitude and make comparisons with objects closer to our environment in the solar system. The total power emitted in the form of gravitational waves is in the order of $7,35\ 10^{24}$W, or 2% of the light energy emitted by the Sun, but because of its remoteness, at a distance of $21,000$ light years, the energy received on Earth is very low: $1,5\ 10^{-17}$ W/m^2 compared to that coming from the Sun ~ 1.4 W/m^2. The double pulsar is in the process of fusion, or coalescence, within a relatively short time on the cosmological scale: about 300 million years.

Since this discovery, a second double pulsar was detected, PSR J0737-3039 was identified in 2004.[15] Its orbit is extremely tight, making it an even richer test source for general relativity. The binary system that enabled the most accurate verification was PSR J1713+0747, a pulsar and white dwarf system discovered in 1993 by R. S. Foster.[16]

[15] Lyne AG, Burgay M, Kramer M, Possenti A, Manchester RN, Camilo F, McLaughlin MA, Lorimer DR, D'Amico N, Joshi BC, Reynolds J, Freire PC. *A double-pulsar system: a rare laboratory for relativistic gravity and plasma physics.* Science. 2004 Feb 20;303(5661):1153–7.

[16] Foster, R. S., Wolszczan, A., Camilo, F., ApJ, 410, L91, (1993).

Chapter 9
Feynman's Path Integrals in Quantum Mechanics

> *You can never solve a problem*
> *on the level on which is was created.*
> **Albert Einstein**

Richard Feynman (1918–1988) is probably the most brilliant theoretical physicist of the second half of 20th century. In his work at Princeton under the direction of John Archibald Wheeler, Feynman sought to solve the problem of divergent expressions in quantum field theory,[1] which, together with Julian Schwinger and Sin-Itiro Tomonaga, was the reason of their award of the 1965 Nobel Prize in Physics "for their fundamental work in quantum electrodynamics, with deep-ploughing consequences for the physics of elementary particles", This has had profound consequences for particle physics. The theory of the *Renormalization Group* has since revealed a depth that guides the theoretical physics of contemporary elementary particles.

In the mean time, in 1941, during his research, he discovered, by talking one evening with Herbert Jehle,[2] a 1932 text by Dirac[3] where he finds a remarkable idea for his research and which, moreover, will allow him to construct a variational formulation, completely new, of non relativistic quantum mechanics[4] In this work, he introduces the mathematical concept of path integrals, which has been developed ever since. This has been the subject of his thesis *Space-Time approach to Non-*

---

[1] John Archibald Wheeler and Richard Phillips Feynman, *Interaction with the Absorber as the Mechanism of Radiation*, Rev. Mod. Phys. 17, 157, 1945.

[2] At a beer party in the Nassau Tavern at Princeton. The account of this event can be found in Feynman's Nobel Conference, https://www.nobelprize.org/prizes/physics/1965/feynman/lecture/.

[3] P.A.M. Dirac, Physikalische Zeitschrift der Sowjetunion 3, No. 1 (1933).

[4] R. P. Feynman, *The principle of least action in quantum mechanics*, Ph. D. thesis, Princeton University, 1942; *Space-Time approach to Non-Relativistic Quantum Mechanics*, Rev. Mod. Phys, vol. 20, p. 367, 1948: see also [26].

© The Author(s), under exclusive license to Springer Nature Switzerland AG 2023
J.-L. Basdevant, *Variational Principles in Physics*,
https://doi.org/10.1007/978-3-031-21692-3_9

185

Relativistic Quantum Mechanics, defended in May 1942 and published only after the end of the war.

9.1   The Initial Click

Let us stand next to Feynman and Jehle the next day, when they open Dirac's 1932 article. An astonishing sentence appears:

> At two times in the vicinity of (t) and $(t + \epsilon)$, the basic transition amplitude $\langle q_2, (t + \epsilon)|q_1, t \rangle$ is analogous to $exp(iS[x]/\hbar)$.

$S[x]$ is the classical expression of the action, but what does *analogous* mean?[5]

Note: A state $|q_i, t\rangle$ is, in Dirac's article, an eigenstate of the position operator $\hat{q}(t)$ in the Heisenberg representation.

We work here in one space dimension, with the spatial variable x. We will see later that everything is easily generalized. The action is, depending on the Lagrangian $\mathcal{L}$

$$S = \int_t^{t+\epsilon} \mathcal{L}(x, \dot{x})dt = \int_t^{t+\epsilon} \left((m\dot{x}^2/2 - V(x)\right) dt \tag{9.1}$$

In traditional terms, the evolution of a state is given by the Schrödinger equation $i\hbar(d/dt)|\psi(t)\rangle = \hat{H}|\psi(t)\rangle$, $\hat{H}$ being the Hamiltonian. If the Hamiltonian does not depend on time, the evolution of a state is $|\psi(t')\rangle = exp(-\frac{i}{\hbar}\hat{H}(t' - t))|\psi(t)\rangle$. The *propagator*[6] $K(\psi', \psi)$ is therefore

$$K(\psi', \psi) = \langle \psi'| \exp(-\frac{i}{\hbar}\hat{H}(t' - t))|\psi\rangle. \tag{9.2}$$

It allows to calculate the transition amplitude between two states.

After a few exchanges, Feynman said to Jehle, let's see if Dirac's "analogous" doesn't simply mean "proportional", that is, whether the infinitesimal evolution over time of a wave function $\psi(x, t)$ cannot be put in the form

$$\psi(x, t + \epsilon) = A \int_{-\infty}^{+\infty} dy \, \exp\left(\frac{i}{\hbar}S[q(t)]\right)\psi(y, t) \tag{9.3}$$

where we integrate on the values of y that is to say the full extent of the initial wave function. The number A is here a quantity to be determined.

---

[5] The exact and tasty words of the dialogue are reported by Feynman in his Nobel Prize-winning address.

[6] Common term due to Feynman, in field theory, for a Green's function.

The action value (9.1) is here

$$S = \epsilon \mathcal{L}(\frac{x-y}{\epsilon}, x) = \epsilon \left(\frac{m}{2}(\frac{x-y}{\epsilon})^2 - V(x) \right) \tag{9.4}$$

In this expression, if the kinetic term is no problem, as regards the potential, Feynman writes, in his thesis and in his 1948 article, directly the above formula. In their book, Feynman and Hibbs use an intermediate value $V(\frac{x+y}{2})$ but, as we will see later in the calculation, the Gaussian integration implies that x and y are very close and this does not matter.

In the integral (9.3), let's set $y = x + \eta$, so we have

$$\psi(x, t + \epsilon) = A \int_{-\infty}^{+\infty} d\eta \exp \left[\frac{i}{\hbar} \frac{m\eta^2}{2\epsilon} - \frac{i\epsilon}{\hbar} V(x) \right] \psi(x + \eta, t). \tag{9.5}$$

In the exponential, the first term $(i/\hbar)(m\eta^2/2\epsilon)$ becomes very large as soon as y is a little different from x. The phase of the integration of (9.5) then varies quickly and this integration oscillates rapidly. The phase of the integrand of (9.5) varies rapidly, this integrand oscillates quickly. On the average, these contributions to the integral cancel each other and only the sufficiently small values of $|x - y| = |\eta|$ contribute appreciably. We can therefore rewrite Eq. (9.9) keeping in mind that only the weak values of $|\eta|$ contribute.

$$\psi(x, t + \epsilon) = A \int_{-\infty}^{+\infty} d\eta \exp \left[\frac{i}{\hbar} \frac{m\eta^2}{2\epsilon} \right] \exp \left[-\frac{i\epsilon}{\hbar} V(x + \eta/2) \right] \psi(x + \eta, t). \tag{9.6}$$

Let us expand, in this equation, the function ψ in series in ϵ and η. By developing the first order in ϵ and the second order in η, we get

$$\psi(x, t + \epsilon) = A \int_{-\infty}^{+\infty} d\eta \exp \left[\frac{i}{\hbar} \frac{m\eta^2}{2\epsilon} \right] \exp \left[-\frac{i\epsilon}{\hbar} V(x + \eta/2) \right] \psi(x + \eta, t). \tag{9.7}$$

The term in $\partial\psi/\partial x$ cancels (odd factor in η). The values of the two remaining Gaussian integrals are recalled in the (9.12) table:

$$C_1 = \int d\eta \exp [\frac{i}{\hbar}(m\frac{\eta^2}{2\epsilon})] = \sqrt{\frac{2i\pi\hbar\epsilon}{m}} \tag{9.8}$$

$$C_2 = \int d\eta \, \eta^2 \exp [\frac{i}{\hbar}(m\frac{\eta^2}{2\epsilon})] = \frac{i\hbar\epsilon}{m} C_1 \tag{9.9}$$

Therefore, the value A of (9.3) is actually a constant, independent of x, equal to

$$A = \frac{1}{C_1} = \sqrt{\frac{m}{2i\pi\hbar\epsilon}}, \tag{9.10}$$

and Feynman finds that he gets the Schrödinger equation:

$$\frac{\hbar}{i}\frac{\partial}{\partial t}\psi(x,t) = -\frac{\hbar^2}{2m}\frac{\partial^2}{\partial x^2}\psi(x,t) + V(x)\psi(x,t). \tag{9.11}$$

Silvan Schweber says that in the autumn of 1946, during the bicentennial of Princeton University, Feynman met Dirac and that the following exchange, rather concise, took place:

Feynman: Did you know that these two quantities are proportional?

Dirac: Are they?

Feynman: Yes.

Dirac: Oh! That's interesting.[7]

Dirac, like Schrödinger and Louis de Broglie, had reread Hamilton's articles, and in particular meditated on the characteristic function and connection between geometric optics and classical mechanics. Dirac was interested in the phase and ratio of the action and the universal Planck constant $\hbar$, in the expression $\exp(i/\hbar)S$, similar to the characteristic function of Hamilton in optics (Chap. 5, Eq. 5.3). But Dirac had not been able to pursue his effort to find the normalization coefficient of this expression. This normalization coefficient is difficult to find, but is essential, as we shall see. Dirac had meanwhile reconsidered the subject by quoting Feynman.[8] But there was a crucial point in Feynman's analysis, which we shall explain. The difficulty, and Feynman's discovery, is that a probability amplitude must be calculated as the sum of amplitudes from all possible paths to get from one point to another, and not just from the classical path of minimal action. At the mathematical level, it is therefore a matter of summing an infinite number of trajectories in what he called a *Path Integral*.

Gaussian Integrals

$$\int_{-\infty}^{\infty} e^{-\lambda x^2}\,dx = \sqrt{\frac{\pi}{\lambda}}, \quad \int_{-\infty}^{\infty} e^{i\lambda x^2}\,dx = \sqrt{\frac{i\pi}{\lambda}}, \quad \int_{-\infty}^{\infty} x^2 e^{i\lambda x^2}\,dx = \frac{1}{2}\sqrt{\frac{-i\pi}{\lambda^3}} \tag{9.12}$$

9.2 Feynman's Principle

Starting with this discovery, Feynman builds a version of quantum mechanics on the following bases.

---

[7] This dialogue is narrated by Silvan S. Schweber, *Feynman's visualization of space-time processes*, Review of Modern Physics, vol. 58, no. 2, 1 April 1986, pp. 449–508.

[8] P. A. M. Dirac, *On the Analogy Between Classical and Quantum Mechanics* Rev. Mod. Phys. 17, 195, 1945.

Feynman's fundamental idea is that a probability amplitude must be calculated as the sum of amplitudes from all possible paths to get from one point to another, and not just from the classical path of minimal action. This consists in generalizing Young's double slit experiment to any process. It will therefore be mathematically necessary to sum the role of an infinite number of trajectories in what he called a *path integral*.

In addition, it states that the amplitude corresponding to a given path is determined by the extension of Hamilton's characteristic function to the action on that path divided by Planck's constant. The total amplitude is of course normalized in accordance with the law of probability that it determines.

What are the main ideas of Feynman, (and Dirac before), in this approach?

1. First of all, both Feynman and Dirac consider that the concept of lagrangian is deeper than that of hamiltonian, even though the latter allows to obtain dazzling results in classic mechanics.

    – The Lagrangian is expressed in the space of configurations, with *space-time* variables, coordinates and their velocities, time derivatives, and quantities that are directly accessible and intuitive, unlike many aspects of Hilbert space and the associated observables. Useful relationships, quantization, uncertainty relationships, scattering amplitudes, without traditional axiomatic tools can be proven directly.
    – An immediate consequence is that it is directly possible to deal with relativistic problems without any major difficulty, contrary to what is known about Hamiltonian formalism. The Lagrangian density is a Lorentz scalar, while the Hamiltonian, fourth component of a four-vector, changes when moving from one referential to another. This transformation property is by no means apparent in traditional quantum mechanics.
    Later on he will extend these considerations to the quantum theory of fields. It is because Feynman poses the problem of quantum mechanics in the *space-time* of Poincaré and Minkowski, that path integrals have been so prolific in quantum field theory.
    – Feynman shows that the transition amplitudes calculated by means of path integrals allow a formulation more efficient than the concept of wave function, especially when dealing with several particles in mutual interaction and perturbation theory. Later he will extend these considerations to quantum field theory.

2. Feynman uses two axioms: a prime quantity, *Probability amplitudes* of processes and its basic property, measurable in quantum mechanics: *the Superposition Principle*.
   In traditional quantum mechanics, of Heisenberg, Schrödinger and Dirac, we must add other postulates: observables and their algebra, the time evolution equation. These notions are consequences in Feynman's approach.

3. It is based on the formula linking the classic action and the transition probabilities (9.3), which we will review in a generalized way, to calculate the transition

probabilities. In other words, it adapts the characteristic function of William Hamilton to quantum mechanics.
4. Finally, and we will see this, classical mechanics appears naturally and in a perfectly continuous way when the orders of magnitude considered are such that the Planck constant becomes negligible. There is, thus, a single physical theory of mechanics, classical and quantum, as Hamilton imagined, and classical mechanics appears as the limit of quantum mechanics when the constant $\hbar$ disappears as being negligible, just as geometric optics appears for short wavelengths.

On the pedagogy of elementary quantum mechanics, Feynman acknowledged that the traditional approach is more efficient. Nevertheless, it is important to present, in this last chapter, the elements of Feynman's approach and the technique of path integrals. We will see the unifying aesthetics that it presents, after the previous chapters.

The book by Feynman and Hibbs *Quantum Mechanics and Path Integrals*, [25], which is a remarkably written teaching manual, proceeds, at each step, by studying a series of particular cases, or exercises. This method allows the reader to gradually absorb the basic ideas of the theory. Here we remain at an elementary level while revealing the structure of the theory and its relationship with analytical mechanics. This is why we will not deal, or only a little, with examples, including the application to perturbation theory, which is remarkably powerful and elegant.

This approach to quantum mechanics was the subject of Feynman's thesis at Princeton in 1942[9] but published after the war in 1948[10] followed a few more years later by the book by Feynman and Hibbs, which for a time constituted his course in quantum mechanics at Caltech, where one will find the essence (and the florilege) of his ideas. Feynman shows that the transitions calculated by means of path integrals allow a formulation more direct than the concept of wave function, especially when dealing with several particles in mutual interaction and perturbation theory. Later he will extend these considerations to quantum field theory.

Countless results have been achieved and this tool plays a central role in contemporary field theory as in many other fields including Statistical Physics (See [27]). The mathematical tool of path integrals with its generalization to quantum field theory has in particular made it possible to carry out the quantization of non-abelian gauge theories in a simpler way than the canonical quantization procedure. This is of great importance in elementary particle theory.

Schwinger (1918–1994) is also the author of a Lagrangian formulation[11] of well-known and renowned quantum mechanics [28]. The results of the two methods are similar, if not identical. However, what clearly differentiates them is the mathematical formulation. Schwinger was infinitely more traditional in his mathematical methods

---

[9] *Feynman's Thesis; A new approach to Quantum Theory: The Principle of Least Action in Quantum Mechanics*, Laurie M. Brown (Editor), World Scientific, Singapore, (2005).

[10] R. P. Feynman emphSpace-Time Approach to Non-relativistic Quantum Mechanics, Rev. Mod. Phys. 20, 367 (1948).

[11] J. Schwinger, Phys. Rev. 82, 914 (1951).

and presented his statements in a differential form, easier to read than Feynman's unconventional formulation.

9.3 The Path Integral

9.3.1 Recollections of Analytical Mechanics

It is useful to recall a few results obtained in analytical mechanics in the previous chapters. In general, we consider one space dimension for simplicity.

1. A mechanical system is characterized by a Lagrangian $\mathcal{L}(x, \dot{x}, t)$, which depends on the state variables (i.e., the position x and its time derivative $\dot{x} = dx/dt$), and possibly on time.
2. The Lagrangian of a particle in a potential $V(x, t)$ is $\mathcal{L} = m\dot{x}^2/2 - V(x, t)$.
3. For *any trajectory* $x(t)$ one can imagine, the *action S* is defined by the integral

$$S = \int_{t_1}^{t_2} \mathcal{L}(x, \dot{x}, t)\, dt. \tag{9.13}$$

4. The least action principle states that the actual physical trajectory $X(t)$ renders S minimal (extremal).
5. The equation of motion that determines the actual trajectory is the Lagrange–Euler equation

$$\frac{\partial \mathcal{L}}{\partial x} = \frac{d}{dt}\left(\frac{\partial \mathcal{L}}{\partial \dot{x}}\right). \tag{9.14}$$

6. For a free particle, $\mathcal{L} = m\dot{x}^2/2$, the classical action between (x_1, t_1) and (x_2, t_2) is

$$S_{cl} = \frac{m}{2}\frac{(x_2 - x_1)^2}{t_2 - t_1}. \tag{9.15}$$

7. If we express the action in terms of the coordinates, the conjugate momenta p_i and the Hamiltonian H are

$$p_i = \frac{\partial S}{\partial x_i}, \quad H = -\frac{\partial S}{\partial t}. \tag{9.16}$$

9.3.2 Quantum Amplitudes

The basic concept on which Feynman relies is that of the *amplitude* of a process. The concept of the *quantum state* of a system (i.e., the description of the state of a system) only comes afterwards. This point of view is more realistic in the sense that

Fig. 9.1 Successive Young
interferences across a series
of screens (only a subset of
possible paths are
represented)

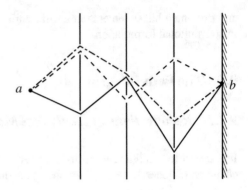

any experiment consists of a series of *processes*. Feynman wants to obtain the laws
of quantum processes. Therefore, Feynman's principle concerns the *dynamics* and,
to a lesser extent, the physical quantities.

Feynman's approach relies on the *superposition principle*. To each physical process, there corresponds complex *amplitudes*, which we denote by ϕ_k, that *add up*.
The probability of observing an event coming from several *interfering alternatives*
for a process is given by the modulus squared of the sum of amplitudes that lead to
this event.

In the Young slit interference experiment, the interfering alternatives correspond
to the passage through each slit. To each of them there corresponds an amplitude
(i.e., ϕ_1 and ϕ_2), and the probability P of observing the outgoing particle at a given
point of the detector is the modulus squared of the sum $P = |\phi_1 + \phi_2|^2$.

We can generalize the experiment by placing a series of screens one after the other,
each of which bears a set of slits. To each possible path of the particle, between the
source and the detector there corresponds an amplitude. The sum of these amplitudes
gives the total amplitude on the detector, and its modulus squared is the probability.

9.3.3 Superposition Principle and Feynman's Principle

Consider a simple process where a particle moves from a point (x_a, t_a) to another
point (x_b, t_b) (we work in space-time, and we include the time variable in the definition of the position of the particle). As in classical physics, one can imagine a
variety of *paths* by which this process can happen. Of course, the classical trajectory
is well defined, and it corresponds to an extremum of the action $S(b, a)$. In quantum
mechanics, all paths $x(t)$ coming from a and ending at b contribute to the amplitude
of the process, as one can visualize from the successive interferences represented in
Fig. 9.1.

The modulus of all these amplitudes is roughly the same,[12] but the *phase* differs appreciably from one path to another. The amplitude $K(b, a)$ is the sum of individual amplitudes

$$K(b, a) = \sum_k \phi_k(b, a), \tag{9.17}$$

where $\phi_k(b, a)$ is the amplitude corresponding to path k.

Feynman writes this sum in the equivalent form

$$K(b, a) = \sum_{\text{all paths } a \to b} \phi(x(t)), \tag{9.18}$$

where $x(t)$ defines a path between a and b. Of course, the specific structure of the setup in Fig. 9.1 does not matter.

The *Feynman principle* consists in stating that, in full generality, in any experimental setup, the phase of the amplitude $\phi(x(t))$ corresponding to a given path is the *classical action* along this path, calculated according to Eq. (9.13), divided by $\hbar$:

$$\phi(x(t)) = \frac{1}{C} e^{\frac{i}{\hbar} S(x(t))}. \tag{9.19}$$

We shall see later on how one fixes the normalization constant C (which is essential). We insist on the fact that the quantity $S(x(t))$ in this expression is the value of the action (9.13) along the path $x(t)$. It does not necessarily correspond to an extremum of the classical action.

9.3.4 Path Integral

This leads us to a central point, which is the evaluation of the sum (9.18) with the definition (9.19). In fact, the family of possible trajectories $x(t)$ between a and b is a complicated set. The result does not correspond to a simple limit of the discrete set, which we could calculate in the case of Fig. 9.1, to a continuous set.

In order to define the sum on all paths, we proceed by first taking discrete time intervals $t_b - t_a$ in the form of N successive equal intervals t_i, $i = 0, \ldots, N$ as:

$$t_b - t_a = N\epsilon, \quad \epsilon = t_{i+1} - t_i \quad t_0 = t_a, \quad t_N = t_b. \tag{9.20}$$

For each value t_i, we choose a value x_i of the variable x. This gives a set of $N - 1$ values since the endpoints are fixed,

$$x_0 = x_a, \quad x_N = x_b.$$

---

[12] Of course, it is only after we have understood the physical and mathematical structure of the problem that this claim appears justified in good approximation.

Fig. 9.2 Example of a
trajectory $x(t)$ in a
discretization of the time
variable. Finding the path
integral consists of
integrating on all values
$x_i(t_i)$ and then taking the
limit $\epsilon \to 0$

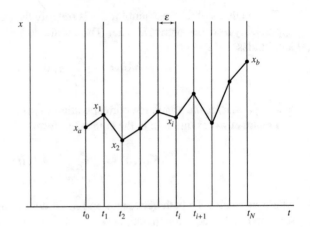

By joining the successive x_i's by straight lines (we shall come back to this point),
we define a trajectory in the form of a broken line that joins the points a and b. Each
set $\{x_i\}$ defines a different possible trajectory.

If we integrate on the values of each x_i from $-\infty$ to $+\infty$, we sum over all
"trajectories" corresponding to this particular discretization of the time variable.
This procedure is illustrated in Fig. 9.2.

For a given value of ϵ, let $C(\epsilon)$ be the normalization constant of (9.19). The
amplitude $K_\epsilon(b, a)$ is given by

$$K_\epsilon(b, a) = \lim_{\epsilon \to 0} \frac{1}{C(\epsilon)} \int \int \cdots \int e^{\frac{i}{\hbar} S(b,a)} \frac{dx_1}{C(\epsilon)} \frac{dx_2}{C(\epsilon)} \cdots \frac{dx_{(N-1)}}{C(\epsilon)}, \qquad (9.21)$$

where, for each value of the set $\{x_i(t_i)\}$, $S(b, a)$ is the action calculated on the
trajectory defined by this set, as represented in Fig. 9.2.

The end of the calculation is to take the limit $\epsilon \to 0$. This is where the normalization factor $C(\epsilon)$ enters, as well as the number of such factors. Indeed, the limit must
exist and only involve physical quantities. Assuming this is achieved, the amplitude
$K(b, a)$ is given by

$$K(b, a) = \lim_{\epsilon \to 0} K_\epsilon(b, a). \qquad (9.22)$$

At this point, the following remarks are in order.

1. Instead of choosing straight lines to join two successive points $x_i(t_i)$ and
   $x_{i+1}(t_{i+1})$, we can perfectly well make the more elegant choice of portions of classical trajectories, which correspond to stationary values of the action $S(i + 1, i)$.
   For a free particle, there is no difference. In the presence of forces, the limit
   $\epsilon \to 0$, which is taken at the end of the calculation, is such that this, also, makes
   no difference in the final result.

2. In this discretization, the value of x_i is well defined, as is its derivative $\dot{x}_i =
   (x_{i+1} - x_i)/\epsilon$. We see, however, that this latter expression is discontinuous and

that the second derivative is not defined at the instants t_i. In the case we consider, this has no importance since the Lagrangian does not involve $\ddot{x}(t)$. In other cases that one can imagine and that are not too pathological, the prescription $\ddot{x} = (x_{i+1} - 2x_i + x_{i-1})/\epsilon^2$ leads to acceptable results.

3. From the mathematical viewpoint, a satisfactory definition of the path integral requires a formalism and some concepts more subtle than this discretization of time. However, for what concerns us here, the only important points are that the summation exists and that the prescriptions (9.21) and (9.22) lead to correct results.

Most of the time in what follows, we will avoid writing the sum over paths as the limits (9.21) and (9.22). We will write this sum as

$$K(b, a) = \int_a^b e^{\frac{i}{\hbar} S(b,a)} \mathcal{D}x(t), \tag{9.23}$$

where the symbol $\mathcal{D}x(t)$ characterizes the mathematical nature of this expression.

The form (9.23) is called a *path integral*. In this expression, $S(b, a)$ is a *number* whose value depends on the *function $x(t)$*. The "integration" over this function $x(t)$, which is represented by $\mathcal{D}x(t)$, is called a *functional integral*.

9.3.5 Amplitude of Successive Events

One justification of the form (9.21) is obtained if we consider the combination law for amplitudes of successive events.

Consider a process $(a \to b)$ and some intermediate time t_c such that $t_a \le t_c \le t_b$. The action $S(b, a)$ is therefore the sum

$$S(b, a) = S(b, c) + S(c, a). \tag{9.24}$$

Indeed, the action is a time integral, and we work with Lagrangians $\mathcal{L}(x, \dot{x}, t)$ that do not depend on higher-order derivatives such as $\ddot{x}$. The integral (9.23) is written as

$$K(b, a) = \int_a^b \exp\left[\frac{i}{\hbar}(S(b, c) + S(c, a))\right] \mathcal{D}x(t),$$

and the previous expression can be rewritten as

$$K(b, a) = \int dx_c \int_c^b \exp\left(\frac{i}{\hbar} S(b, c)\right) \mathcal{D}x(t) \int_a^c \exp\left(\frac{i}{\hbar} S(c, a)\right) \mathcal{D}x(t), \tag{9.25}$$

where, of course, the integral over x_c is a usual integral. We have factorized the integrand.

This expression is a usual integral,

$$K(b, a) = \int dx_c K(b, c) K(c, a). \tag{9.26}$$

In other words, the amplitudes for two successive events going through the same given intermediate point c, $(a \to c)$ and $(c \to b)$, are *multiplied*. The amplitude $K(b, a)$ is the *sum* of these products on all possible values of the intermediate point. This is simply the *superposition principle*, which is a basic principle of Feynman.

This argument can be extended to any number N of intervals, with intermediate points x_i, $i = 1, \cdots (N - 1)$, which leads to

$$K(b, a) = \int K(b, N - 1) K(N - 1, N - 2) \cdots K(i + 1, i)$$
$$\cdots K(1, a) dx_1 \, dx_2 \ldots dx_{N-1}. \tag{9.27}$$

Assuming these intervals are infinitesimal and of equal length ϵ, the corresponding expression resembles Eq. (9.21). It is not identical, however, since the latter form is a limit, whereas (9.27) is an equality. This, however, enables us to obtain an infinitesimal form of the amplitude K between two points separated by an infinitesimal time interval ϵ. In fact, when $t_2 - t_1 = \epsilon$ is infinitesimal, the action (9.13) is, to first order in ϵ,

$$S(2, 1) = \epsilon \mathcal{L} \left(\frac{x_2 + x_1}{2}, \frac{x_2 - x_1}{\epsilon}, \frac{t_2 + t_1}{2} \right), \tag{9.28}$$

or

$$K(2, 1) = \frac{1}{C(t_2 - t_1 = \epsilon)} \exp \left(\frac{i\epsilon}{\hbar} \mathcal{L} \left(\frac{x_2 + x_1}{2}, \frac{x_2 - x_1}{\epsilon}, \frac{t_2 + t_1}{2} \right) \right). \tag{9.29}$$

Inserting this result into the formula (9.27), and assuming we can exchange the order of the integration and the limit $\epsilon \to 0$, we indeed obtain an equality between the two expressions. This justifies the method (9.21) and (9.22) in all cases where the expressions converge sufficiently well.

9.4   Free Particle

We first apply what we have done to the case of a particle propagating freely in space. One calls *propagator* the amplitude $K(b, a)$ of the propagation of a particle (free or not) from point a to point b.

A free classical particle propagates according to a uniform linear motion. The corresponding classical action is

$$S_{cl} = \frac{m}{2} \frac{(x_2 - x_1)^2}{t_2 - t_1}. \tag{9.30}$$

This result is obvious. In this problem, the velocity $\dot{x}$ is a constant and the Lagrangian is $\mathcal{L} = m\dot{x}^2/2 = m[(x_2 - x_1)/(t_2 - t_1)]^2/2$, which leads directly to (9.30) since $S = \int \mathcal{L} \, dt$.

9.4.1 Propagator of a Free Particle

In order to calculate the propagator of a free particle, we could use the limiting form (9.21). However, in this case, it is completely equivalent to use the expression (9.27) because the result is independent of the value of $\epsilon = t_{i+1} - t_i$. We will also obtain the value of the normalization coefficient $C(\epsilon)$ in (9.19).

The final result is that the propagator of a free particle between points a and b is

$$K(b, a) = \left(\frac{2\pi i \hbar (t_b - t_a)}{m} \right)^{-1/2} \exp \left(\frac{im(x_b - x_a)^2}{2\hbar(t_b - t_a)} \right). \tag{9.31}$$

This gives the value of the normalization factor. For a free particle and a time interval $(t_b - t_a)$, we read in the formula above

$$C(t_b - t_a) = \left(\frac{2\pi i \hbar (t_b - t_a)}{m} \right)^{1/2}. \tag{9.32}$$

The normalization constant C depends only on the variable $(t_b - t_a)$.

For an infinitesimal time interval $(t_{i+1} - t_i) = \epsilon$, which is the case in the definition (9.21) of the path integral, we have

$$C(\epsilon = t_{i+1} - t_i) = \left(\frac{2i\pi\hbar\epsilon}{m} \right)^{1/2}. \tag{9.33}$$

This prescription ensures that the path integral (9.22) exists.

The proof of this is the following. (A complete mathematical analysis can be found in [29]).

In the expression (9.27) we set $x_0 \equiv x_a$. We first calculate the integral over x_1. Using (9.29), we obtain

$$K(2, 0) = \left(\frac{1}{C(\epsilon)} \right)^2 \int dx_1 \exp \left(\frac{im}{2\hbar\epsilon} \left[(x_2 - x_1)^2 + (x_1 - x_0)^2 \right] \right). \tag{9.34}$$

With the identity

$$(x_2 - x_1)^2 + (x_1 - x_0)^2 = \frac{1}{2}\left((x_2 + x_0)^2 + 4\left(x_1 - \frac{x_2 - x_0}{2}\right)^2\right)$$

this reduces to a simple Gaussian integral on x_1, which gives

$$K(2, a) = \left(\frac{1}{C(\epsilon)}\right)^2 \sqrt{\frac{i(2\epsilon)\pi\hbar}{m}} \exp\left(\frac{im}{4\hbar\epsilon}(x_2 - x_0)^2\right). \tag{9.35}$$

Of course, we have made use of the fact that $t_2 - t_1 = t_1 - t_0 = \epsilon$, but we have not taken any limit.

The expression (9.35) can be written in terms of $(t_2 - t_0) = 2\epsilon$ as

$$K(2, a) = \left(\frac{1}{C(\epsilon)}\right)^2 \sqrt{\frac{i(2\epsilon)\pi\hbar}{m}} \exp\left(\frac{im}{2\hbar(t_2 - t_a)}(x_2 - x_0)^2\right). \tag{9.36}$$

To first order in ϵ, this expression must be the same as that of $K(1, a)$ if $t_1 - t_a = \eta = 2\epsilon$ and $x_1 = x_2$, and therefore

$$K(1, a)_{t_1 - t_a = 2\epsilon} = \frac{1}{C(2\epsilon)} \exp\left(\frac{im}{2\hbar(t_2 - t_a)}(x_2 - x_0)^2\right). \tag{9.37}$$

Consequently, the equality of (9.37) and (9.36) for infinitesimal time intervals $t_b - t_a = \epsilon$ imposes the choice

$$C(\epsilon) = \left(\frac{2i\pi\hbar\epsilon}{m}\right)^{1/2} \equiv \left(\frac{2i\pi\hbar(t_b - t_a)}{m}\right)^{1/2}. \tag{9.38}$$

The proof of (9.31) can be obtained by recursion. Assuming this result is exact, we insert it in (9.26) by considering an intermediate point (x, t) we obtain

$$K(b, a) = \left(\frac{2\pi i\hbar(t_b - t)}{m} \frac{2\pi i\hbar(t - t_a)}{m}\right)^{-1/2}$$
$$\int \exp\left(\frac{im}{2\hbar}\left(\frac{(x_b - x)^2}{(t_b - t)} + \frac{(x - x_a)^2}{(t - t_a)}\right)\right) dx. \tag{9.39}$$

It is straightforward to obtain

$$\frac{(x_b - x)^2}{(t_b - t)} + \frac{(x - x_a)^2}{(t - t_a)} =$$
$$\frac{(x_b - x_a)^2}{(t_b - t_a)} + \frac{(t_b - t_a)}{(t_b - t)(t - t_a)}\left(x - \frac{x_b(t - t_a) + x_a(t_b - t)}{(t_b - t_a)}\right)^2. \tag{9.40}$$

The first term, which is independent of x, factorizes in the integral, which boils down to a Gaussian integral. The value I of this integral (without the prefactors of (9.39)) is therefore

$$I = \left(\frac{m(t_b - t_a)}{2i\pi\hbar(t_b - t)(t - t_a)} \right)^{-1/2} \exp\left(\frac{im}{2\hbar} \frac{(x_b - x_a)^2}{(t_b - t_a)} \right). \qquad (9.41)$$

Inserting this in (9.39), we obtain the expected result

$$K(b, a) = \left(\frac{2\pi i\hbar(t_b - t_a)}{m} \right)^{-1/2} \exp\left(\frac{im}{2\hbar} \frac{(x_b - x_a)^2}{(t_b - t_a)} \right). \qquad (9.42)$$

The proof is completed by noticing that the previous result does not require any condition on a, b and the intermediate point (x, t). Therefore, the method can be extended to any partition of the interval $[a, b]$. The formula therefore coincides with the "definition" (9.35) in the infinitesimal limit $(t_b - t_a) = N\epsilon$ if it is legitimate to interchange the order of the integration and the limit $\epsilon \to 0$.

Free propagator in Schrödinger theory

In Schrödinger's theory, the hamiltonian of a free particle is $\hat{H} = \hat{p}^2/2m$ and does not depend on x. The free propagator (9.2) is then diagonal in the basis $|p\rangle$ of the *momentum* eigenstates. We therefore have

$$K(x', t'; x, t) = \int dp \, \frac{e^{ipx'/\hbar}}{\sqrt{2\pi\hbar}} \exp\left[-ip^2(t' - t)/(2m\hbar) \right] \frac{e^{-ipx/\hbar}}{\sqrt{2\pi\hbar}}. \qquad (9.43)$$

By rewriting the exponential argument, we have:

$$\frac{ip(x' - x)}{\hbar} - \frac{ip^2(t' - t)}{(2m\hbar)} = \frac{-i(t' - t)}{2m\hbar} \left(p - \frac{m(x' - x)}{t' - t} \right)^2 + \frac{im(x' - x)^2}{2\hbar(t' - t))}$$

so that the expression (9.43) which actually factors into an exponential times a Gaussian integral, is *identical* to (9.42):

$$K(x', t'; x, t) = \sqrt{\frac{m}{2i\pi\hbar(t' - t)}} \exp\left(\frac{im(x' - x)^2}{2\hbar(t' - t)} \right). \qquad (9.44)$$

9.4.2 Evolution Equation of the Free Propagator

For simplicity, let us fix the origin of time and position at a. We call $b \equiv (x, t)$ the point of arrival, and we examine the properties of the propagator $K(x, t; 0, 0)$ as a function of the endpoint (x, t). If we set $\mathcal{K}(x, t) \equiv K(x, t; 0, 0)$, we have

$$\mathcal{K}(x, t) = \sqrt{\frac{m}{2\pi i \hbar t}} \, \exp\left(\frac{im}{2\hbar} \frac{x^2}{t}\right). \tag{9.45}$$

One can check with no difficulty that the free propagator obeys the partial differential equation

$$i\hbar \frac{\partial \mathcal{K}}{\partial t} = -\frac{\hbar^2}{2m} \frac{\partial^2 \mathcal{K}}{\partial x^2} \tag{9.46}$$

for $t > 0$ (or $t_b > t_a$).

Equation (9.46) has the same form as the Schrödinger equation for a free particle. We must, however, be careful since we do not yet know the physical nature of the amplitude K and how it is related to a physical *probability amplitude*.

9.4.3 Normalization and Interpretation of the Propagator

Indeed, if we calculate the integral over x of $\mathcal{K}(x, t)$, we obtain

$$\int_{-\infty}^{+\infty} \mathcal{K}(x, t)\, dx = \sqrt{\frac{m}{2\pi i \hbar t}} \int_{-\infty}^{+\infty} \exp\left(\frac{im}{2\hbar} \frac{x^2}{t}\right) dx = 1 \quad \forall t > 0. \tag{9.47}$$

In particular, in the limit $t = \epsilon \to 0$, we have

$$\lim_{t \to 0} \mathcal{K}(x, t) = \delta(x), \tag{9.48}$$

where δ is the Dirac distribution.

Therefore, K is not strictly speaking a quantum mechanical *probability amplitude*. However, this is a minor problem. In fact, the propagator $K(b, a)$ is the amplitude for a particle going from *point a* to *point b*. It is unphysical in the sense that it is not possible physically to prepare or to measure a particle whose position is strictly defined to be the point $x = x_a$. The position "eigenstates" $|x\rangle$ have "wave functions" $\psi_{x_a}(x) \propto \delta(x - x_a)$ in the same way that the momentum eigenstates have wave functions $\psi_{p_0}(x) \propto \exp(ip_0 x)$. These are not physical since they are not normalizable. They are "eigendistributions" that do not belong to the Hilbert space. Nevertheless, we know that any physical state can be written as a linear superposition of such nonphysical states. It is useful to work with the nonphysical states $|x\rangle$ and $|p\rangle$ in all intermediate calculations.

The same thing happens here. It is convenient to work with the propagator $K(b, a)$ and to call it a *probability amplitude*, even though we are aware that a true probability amplitude is obtained after a summation of $K(b, a)$ over vicinities of b and a.

One can check that if we forget this precaution, the "probability" of observing the particle in a vicinity dx of point x_b, knowing that it originates from a, would be

$$P(b)\, dx = \frac{m}{2\pi\hbar(t_b - t_a)}\, dx,$$

whose integral over all space is infinite. This is exactly the same problem as encountered in quantum mechanics to shift from de Broglie plane waves to wave packets.

9.4.4 Fourier and Schrödinger Equations

Some authors stress the fact that the free Schrödinger equation can be considered as a Fourier diffusion equation,

$$\frac{\partial\rho}{\partial t} = D\nabla^2\rho,$$

for a purely imaginary time $t = i\tau$. This remark is interesting in that the same mathematical techniques apply to both and that the solutions have obvious formal similarities.

Two points are in order. First, the function $\mathcal{K}$ that we use here becomes a density ρ (of heat or matter), which is positive. The solution is then real and positive, or zero. The result (9.47) expresses the conservation of energy, and the limit (9.48) represents an initial condition where some quantity of heat has been deposited on a given point, which avoids any problem of interpretation.

Secondly, and this is perhaps more interesting, this is an example of the fact that path integral techniques are useful in a large category of problems. One can refer to the remarkable book *Techniques and Applications of Path Integration* by Schulman [27]. In the present case, the solution of a partial differential equation of first order in time can be cast quite directly into the form of a path integral. This is the case for the Fourier equation as well as for the Schrödinger equation.

9.4.5 Energy and Momentum of a Free Particle

The propagator $\mathcal{K}(x, t)$ oscillates in x and in t with a wavelength λ and a frequency ω that vary with x. Locally, for large values of x and t, in a region ($\delta X \ll x$, $\delta t \ll t$), one can approximate this wave by a monochromatic plane wave

$$\mathcal{K}(x,t) = \sqrt{\frac{m}{2\pi i\hbar t}} \, \exp i\phi(x,t) \propto e^{i(kx-\omega t)}, \tag{9.49}$$

where k is the wave vector and ω the frequency. These are locally related by

$$k = \frac{\partial\phi}{\partial x} \qquad \omega = -\frac{\partial\phi}{\partial t}. \tag{9.50}$$

Here, the value of the phase is

$$\phi(x,t) = \frac{m}{2\hbar}\frac{x^2}{t}.$$

Therefore, we obtain

$$k = \frac{m}{\hbar}\frac{x}{t}, \quad \text{i.e., } \lambda = \frac{2\pi}{k} = \frac{h}{m(x/t)}, \tag{9.51}$$

and

$$\omega = \frac{m}{2\hbar}\left(\frac{x}{t}\right)^2. \tag{9.52}$$

If we place ourselves far from the origin $x \gg \lambda$, the propagator oscillates in x and t with a wavelength λ and a frequency ω, which are both nearly constant. If the particle, emitted at the origin at $t = 0$ is detected at point x at time t, its velocity is $v = x/t$, its momentum is $p = mv = m(x/t)$, and its kinetic energy is $E_k = m(x/t)^2/2$.[13] We therefore obtain the de Broglie relation between the wavelength of the propagator and the momentum of the free particle

$$\lambda = \frac{h}{p}. \tag{9.53}$$

Similarly, the kinetic energy $E_k = mv^2/2$ of the free particle is related to the frequency ω of the wave by the relation $\omega = (m/2\hbar)(x^2/t^2)$; i.e.,

$$E_k = \hbar\omega. \tag{9.54}$$

These de Broglie–Einstein relations (9.53) and (9.54) will appear later as being in agreement with the definition of energy and momentum in the classical limit.

---

[13] See, for instance, [11], Chap. 2, Sect. 6, for a discussion of this point.

9.4.6 Interference and Diffraction

The calculation of interference and diffraction phenomena follows immediately from Eq. (9.26),

$$K(b, a) = \int dx_c K(b, c) K(c, a). \tag{9.55}$$

In three dimensions, this result is transposed as

$$K(b, a) = \int d^3\mathbf{r}_c K(b, c) K(c, a). \tag{9.56}$$

We can define a function $G(r_c)$ as $G = 1$ in some domain D (collection of slits, aperture of arbitrary shape) and $G = 0$ elsewhere. G represents a screen that lets the particles go through freely in D and stops them elsewhere. The amplitude emitted in a, diffracted by G, and measured on each point b of a screen or detector is simply the usual sum

$$K_G(b, a) = \int d^3\mathbf{r}_c G(\mathbf{r}_c) K(b, c) K(c, a). \tag{9.57}$$

In the book by Feynman and Hibbs, there are several examples and calculations of this type. We shall not elaborate further on this aspect.

9.5 Wave Function and the Schrödinger Equation

Up to now, we have considered the amplitude for a particle to travel from a to b.

It is quite possible, and legitimate, to address the question of the *total* amplitude of a particle to reach an arbitrary point b, independently of what happened to it previously (up to now, we considered the problem when "the particle was emitted at a"). Of course, this amplitude is the *wave function* $\psi(x, t)$[14] (we have replaced the name b by variables (x, t)). This probability amplitude $\psi(x, t)$ obviously satisfies all the conditions we have found previously. By definition, it is square integrable, which avoids by construction the limiting procedures seen in Sect. 9.4.3. Apart from this, the amplitude $\mathcal{K}$ of (9.45) is a particular wave function for which we know that the particle started at $a \equiv (0, 0)$.

The wave function is a probability amplitude. Therefore, it satisfies the law of composition of successive amplitudes (9.26); i.e., the integral equation

$$\psi(x, t) = \int_{-\infty}^{\infty} K(x, t; x', t') \psi(x', t') \, dx'. \tag{9.58}$$

---

[14] See, for instance, [11], Chap. 2, Sect. 1.

The physical content of this formula is important. The amplitude $\psi(x, t)$ for the particle to arrive at (x, t) is the sum over all possible values of an intermediate point x' of the product of the total amplitude $\psi(x', t')$ and the amplitude $K(x, t; x', t')$ to go from (x', t') to (x, t).

In other words (we intentionally keep the enthusiastic presentation of Feynman), the effect of all the past history of a particle is contained in a single function $\psi(x, t)$. One can forget everything one knows about the past history of a particle. If one knows its wave function at a given time t, one can calculate and "read" in it all that can happen to the particle in the future.[15]

In fact, Eq. (9.58) is nothing else than the modern expression of the Huygens–Fresnel principle in optics (see, for instance, Born and Wolf [14], Chap. VIII), which founded wave optics. The Huygens principle, given in 1690, was that "Each infinitesimal element of a wave front can be considered as a secondary perturbation which radiates spherical wavelets. The wave front at a later time is the envelope of these wavelets". Fresnel completed this principle later, in 1818, by postulating that the secondary wavelets "are in mutual interference". The fundamental principles of wave optics were stated.

9.5.1 Free Particle

We have abundantly treated the case of a free particle above. The propagator can be calculated with no difficulty:

$$K(x_2, t_2, x_1, t_1) = \sqrt{\frac{m}{2\pi i \hbar (t_2 - t_1)}} \, \exp\left(\frac{im}{2\hbar} \frac{(x_2 - x_1)^2}{(t_2 - t_1)} \right). \tag{9.59}$$

The wave function $\psi(x, t)$ satisfies the free Schrödinger equation

$$i\hbar \frac{\partial \psi(x, t)}{\partial t} = -\frac{\hbar^2}{2m} \frac{\partial^2 \psi(x, t)}{\partial x^2}. \tag{9.60}$$

As in the case of a particle in a potential, which we examine in the next paragraph, we will not further pursue the analysis, which is completely analogous to the usual analysis of Schrödinger theory. One can refer to the book by Feynman and Hibbs [25] for all details.

---

[15] Feynman added, with his legendary sense of humor, "The effect of the entire History on the future of the universe could be obtained from a single gigantic wave function."

9.5.2 Particle in a Potential

We have already dealt, at the beginning, with the case of the Schrödinger equation for a particle placed in a potential $V(x)$. We can easily consider a potential $V(x, t)$ that varies with time (it's done in the book of Feynman and Hibbs [25]).

We will not pursue the analysis of quantum mechanics in general, we can refer to the book of Feynman and Hibbs.

Let us note only the following points.

1. The theory of observables, their algebraic properties, We shall see two important cases: Hamiltonian and momentum.
2. There are significant technical and conceptual simplifications in the design and treatment of perturbation theory.
3. The resulting formalism extends much more easily to several particles.

9.5.3 Hamiltonian Operator and Consequences

It is a simple calculation to extend what we have just done to situations that differ in the number of variables. In three dimensions we know that we are able to find the Schrödinger equation

$$i\hbar\frac{\partial\psi(\mathbf{r}, t)}{\partial t} = -\frac{\hbar^2}{2m}\nabla^2\psi(\mathbf{r}, t) + V(\mathbf{r})\psi(\mathbf{r}, t). \tag{9.61}$$

These equations are written in the form

$$i\hbar\frac{\partial\psi}{\partial t} = \hat{H}\psi \tag{9.62}$$

where the Hamiltonian **operator** $\hat{H}$ is

$$\hat{H} = -\frac{\hbar^2}{2m}\nabla^2 + V \tag{9.63}$$

It is of course possible to extend this result to situations where the interaction is different. The same calculation is a little longer in the well-known case of a charge particle q plunged into a magnetic field B and an electric field E deriving from the potentials $\mathbf{A}$, Φ and undergoing the Lorentz force of Lorentz, where the lagrangian is

$$L = m\dot{x}^2/2 + q\,\dot{\mathbf{r}}\cdot\mathbf{A}(\mathbf{r}, t) - q\,\Phi(\mathbf{r}, t). \tag{9.64}$$

The three-dimensional calculation is similar to the previous one, with some complications that are treated by L. S. Schulman ([27] Chap. 4), which refers to what is said by Feynman himself in his original article [26]. It leads to the well-known Schrödinger equation:

$$i\hbar\frac{\partial\psi(\mathbf{r},t)}{\partial t} = \frac{1}{2m}\left(\frac{\hbar}{i}\nabla - q\mathbf{A}\right)\cdot\left(\frac{\hbar}{i}\nabla - q\mathbf{A}\right)\psi + q\Phi\,\psi, \qquad (9.65)$$

therefore to the expression of the Hamiltonian $\hat{H}$

$$\hat{H} = \frac{1}{2m}\left(\frac{\hbar}{i}\nabla - q\mathbf{A}\right)\cdot\left(\frac{\hbar}{i}\nabla - q\mathbf{A}\right) + q\Phi. \qquad (9.66)$$

Let us emphasize that this form comes from the classical Lagrangian and the technique of path integrals. We have not introduced the conjugate momentum $\mathbf{p} = (\hbar/i)\nabla$, whose definition comes later in Feynman's presentation.

9.5.4 Conservation of Probability

The wave equation (9.62) means that if $\psi(x,t)$ is a probability amplitude, then there is conservation of probability.

First we see that if f and g are square integrable, then

$$\int_{-\infty}^{+\infty}(Hg)^*f\,dx = \int_{-\infty}^{+\infty}g^*(Hf)dx \qquad (9.67)$$

Indeed, after an integration by parts,

$$-\frac{\hbar^2}{2m}\int_{-\infty}^{+\infty}\frac{d^2}{dx^2}(g^*)f\,dx = -\frac{\hbar^2}{2m}\int_{-\infty}^{+\infty}g^*\frac{d^2}{dx^2}f\,dx. \qquad (9.68)$$

We thus see that the hamiltonian $\hat{H}$ is *hermitian*.

Next, we note that the ψ^* complex conjugate of ψ obeys a Schrödinger equation where each i is changed to $-i$

$$\frac{\partial\psi^*}{\partial t} = +\frac{i}{\hbar}(\hat{H}\psi^*). \qquad (9.69)$$

Therefore

$$\frac{\partial\psi^*}{\partial t}\psi = -\frac{\partial\psi}{\partial t}\psi^* \qquad (9.70)$$

Hence the conservation of probability

$$\int_{-\infty}^{+\infty} \frac{\partial \psi^*}{\partial t} \psi dx + \int_{-\infty}^{+\infty} \psi^* \frac{\partial \psi}{\partial t} dx = \frac{d}{dt} \left(\int_{-\infty}^{+\infty} \psi^* \psi dx \right) = 0. \qquad (9.71)$$

This result can be obtained directly from the starting hypotheses of Feynman, and his propagator. If f is the wave function at the moment t_a, then the integral of its square modulus squared is the same at time t_b:

$$\text{if} \quad \psi(b) = \int K(b, a) f(a) dx_a, \quad \text{then} \quad \int_{-\infty}^{+\infty} \psi^*(b) \psi(b) dx_b = \int_{-\infty}^{+\infty} \psi^*(a) \psi(a) dx_a$$

which must be true for any f. This property stems from

$$\int_{-\infty}^{+\infty} K^*(b; x_a', t_a) K(b; x_a, t_a) = \delta(x_a' - x_a). \qquad (9.72)$$

9.5.5 Stationary States

We consider here the cases where the Hamiltonian is time independent.

The problem of stationary states, eigenstates of energy, is obviously more direct to treat from the Schrödinger equation (9.62). Here, if we look for a particular solution of the form $\psi(x, t) = \phi(x) f(t)$ where position and time dependences are factored, we get

$$\phi(x) f'(t) = (-i/\hbar)[\hat{H}\phi(x)] f(t)$$

$$\frac{f'(t)}{f(t)} = \frac{-i}{\hbar} \frac{\hat{H}\phi(x))}{\phi(x)}. \qquad (9.73)$$

The equality of two functions, one of x, the other of t, means that they are both equal to a constant, which we set equal to $-(i/\hbar)E$. This particular solution is therefore of the form

$$\psi(x, t) = \phi(x) \exp^{(-i/\hbar)Et} \quad \text{with} \quad \hat{H}\phi(x) = E\phi(x). \qquad (9.74)$$

In particular, this means that the wave function oscillates at any point in space with the same well-defined frequency. We have seen in (9.54) that the frequency at which the phase oscillates corresponds to the energy. The system therefore has a well defined energy.

The presence probability of the particle at a point is given by the modulus squared $|\psi(x)|^2$ of the wave function at that point. This probability is invariant in time, the system is in a stationary state.

Similarly, by analysing the total presence probability of a superposition of energy eigenfunctions, it is easy to deduce that two separate functions ϕ_1, ϕ_2 corresponding to eigenvalues E_1, E_2 are orthogonal

$$\int_{-\infty}^{+\infty} \phi_1^*(x)\phi_2(x)dx = \int_{-\infty}^{+\infty} \phi_1(x)\phi_2^*(x)dx = 0.$$

9.6 The Momentum

By defining the momentum operator, we open the way to all traditional observables, function of positions and momentum.

9.6.1 Momentum Measurement

One way to measure the momentum of a particle (or any other quantity) is to organize a process of position measurements that tells us about this quantity.

Let us consider a one-dimensional particle, originally located, and proceed to the next step.

We will observe where the particle ended up after a time $t = T$. If the position is y then the particle speed is equal to y/T and the amount of motion, or momentum, is $p = my/T$. In fact, to have a precise determination of the law of probability of the momentum, that is the probability that the value of the momentum is between p and $p + dp$, it is necessary to measure the probability $P(y)\,dy$ that the free particle, that is, being disconnected from any interaction, ends up between y and $y + dy$. So we determine p from the measurement of y by $p = my/T$.

Let $f(x)$ be the particle wavefunction at $t = 0$. We want to express $P(p)$ directly from $f(x)$.

The probability amplitude that the particle will arrive at y at time $t = T$ is

$$\psi(y, T) = \int_{-\infty}^{+\infty} K_0(y, T; x, 0) f(x)\, dx. \tag{9.75}$$

If we put the value of the free propagator (9.31) in this expression, we get

$$\psi(y, T) = (\frac{m}{2\pi i\hbar T})^{1/2} \exp\{\frac{imy^2}{2\hbar T}\} \int_{-\infty}^{+\infty} \exp\{\frac{im(-2yx + x^2)}{2\hbar T}\} f(x)\, dx. \tag{9.76}$$

The modulus squared of this expression gives the probability that the particle is between y and $y + dy$ that is, in the limit $T \to \infty$, the probability that the momentum is between p and $p + dp$

$$P(y)dy = \frac{mdy}{2\pi \hbar T} \left| \int_{-\infty}^{+\infty} \exp\{\frac{im(-2yx + x^2)}{2\hbar T}\} f(x)\,dx \right|^2 = P(p)dp \text{ for } T \to \infty. \tag{9.77}$$

Since the presence probability decreases rapidly beyond a certain extension b, it is necessary to choose T such that $imb^2/2\hbar T$ becomes negligible.

By inserting the relationship $p = my/T$ and moving to the limit $T \to \infty$, we thus obtain

$$P(p)dp = \frac{dp}{2\pi \hbar} \left| \int_{-\infty}^{+\infty} \exp\{\frac{-ipx}{\hbar} + \frac{imx^2}{2\hbar T}\} f(x)\,dx \right|^2. \tag{9.78}$$

9.6.2 Probability Amplitude

Starting from a wave function $\psi(x, t)$ we can therefore construct the probability amplitude $\phi(p, t)$ such that $P(p, t) = |\phi(p, t)|^2$.

$$\phi(p, t) = \int_{-\infty}^{+\infty} \exp\{\frac{-ipx}{\hbar}\} \psi(x, t) \frac{dx}{(2\pi \hbar)^{1/2}}. \tag{9.79}$$

This expression extends to three dimensions as:

$$\phi(\mathbf{p}, t) = \int_{-\infty}^{+\infty} \exp\{\frac{-i\mathbf{p} \cdot \mathbf{r}}{\hbar}\} \psi(\mathbf{r}, t) \frac{d\mathbf{r}}{(2\pi \hbar)^{3/2}}. \tag{9.80}$$

9.6.3 Fourier Transformation

The two amplitudes are Fourier transforms of each other. The inverse Fourier transform of (9.79) is

$$\psi(x, t) = \int_{-\infty}^{+\infty} \exp\{\frac{ipx}{\hbar}\} \phi(p, t) \frac{dp}{(2\pi \hbar)^{1/2}}. \tag{9.81}$$

The isometry of the Fourier transformation means that if $f_1(x)$ and $f_2(x)$ are transforms of $g_1(p)$ and $g_2(p)$, then

$$\int_{-\infty}^{+\infty} f_1^*(x) f_2(x)dx = \int_{-\infty}^{+\infty} g_1^*(p)g_2(p)dp, \tag{9.82}$$

so that

$$\int_{-\infty}^{+\infty} |\psi(x,t)|^2 dx = \int_{-\infty}^{+\infty} |\phi(p,t)|^2 dp = 1, \tag{9.83}$$

$\phi(p,t)$ and $\psi(x,t)$ are both probability amplitudes, and either can represent the state of a particle and its evolution.

Feynman and Hibbs present several examples, such as cases of diffraction, where each time the Heisenberg inequality is satisfied. They do not give a proof of the relationship itself, which, if violated, would mark the end of quantum mechanics. This inequality is a very simple consequence of Fourier's analysis for the probability amplitudes of $\psi(x,t)$ and $\phi(p,t)$ that we have just obtained above.

9.7   Concluding Remarks

Both from the conceptual and the technical points of view, the method of Feynman path integrals has an undeniable elegance and richness. We have mentioned that it extends to many other physical problems such as quantum field theory, Brownian motion, polarons, spin physics, statistical mechanics, and critical phenomena, as one can see in the book of Schulmann [27]. This book contains, in particular, a very pleasant discussion of quantum mechanics in curved spaces. We end this chapter with a series of remarks that the present results have induced after going through the previous five chapters of this book.

There is no hierarchical relationship between the depth of the various approaches and different chapters of physics, neither do we wish to discuss any axiomatics of physics. It is a personal matter of taste to prefer such and such a line of thought. What is interesting here is to see the unifying character of what we have discussed, from the Fermat principle up to the Feynman path integrals.

9.7.1   Classical Limit

Consider again the path integral (9.23)

$$K(b,a) = \int_a^b e^{\frac{i}{\hbar} S(b,a)} \mathcal{D}x(t), \tag{9.84}$$

and suppose the classical action $S(b,a)$ is *macroscopic*, i.e., it is much larger than the Planck constant $\hbar$. Consider the contribution of several paths that can perfectly well be close to each other in the classical sense but whose difference is much larger than $\hbar$. The contributions of these paths to the phase will be completely different (and very difficult to determine with an accuracy better than, say, π). With great probability, they will interfere destructively. If one considers the set of all those paths, their total contribution to the integral will vanish.

However, in the vicinity of the classical trajectory $x_{cl}(t)$, the action $S_{cl}(b,a)$ is stationary. Therefore, paths that are sufficiently close to the classical trajectory will give contributions that will interfere in a constructive way. Only those paths along which the action $S(b,a)$ is sufficiently close to the classical action $S_{cl}(b,a)$ will contribute, the difference being noticeably smaller than the unit of action $\hbar$. Notice that for all processes involving macroscopic values of the action, this quantity will be larger than, say, $10^{(25\ to\ 30)}\hbar$.

In other words, under these conditions, the only appreciable contribution will come from an infinitesimal vicinity of the *classical trajectory* that cannot be resolved experimentally. Consequently, the "probability" of the classical trajectory is equal to one. The probability for any trajectory that can be distinguished from the classical one vanishes.

Therefore, classical mechanics appears here as the *limit* of quantum mechanics for macroscopic actions. Of course, one may wonder about the fact that Feynman's starting point involves the classical action in (9.84), which means that some care should be taken with the previous assertion. However, from the very beginning, Feynman operates in *space-time* (x,t). Therefore, all quantities defined in Sect. 9.3.1 (i.e., $(x,\dot{x},t)$, the Lagrangian $\mathcal{L}$, and the action S) are perfectly well-defined quantities, even though they do not have to possess any intuitive meaning.

This approach removes one of the sometimes confusing aspects of traditional quantum mechanics that tends to make us live in abstract spaces of infinite dimension. Consequently, all the quantities defined in Sect. 9.3.1, that is $(x,\dot{x},t)$, the lagrangian $\mathcal{L}$ and the action S, are perfectly defined, even if they would not be allocated any intuitive physical significance.

In addition, in theoretical physics, this approach makes it possible to deal with problems directly in the space-time of Einstein, Lorentz and Poincaré. Special relativity can therefore be easily incorporated into calculations. On this subject, it is interesting to see that Feynman does not hesitate to take historical examples such as the Lamb shift of the levels of the hydrogen atom, which was long considered as one of the summits of relativistic quantum theory.

It is thus understandable that while this approach to quantum mechanics remains on the sidelines of the methods of teaching elementary theory, it has for many years been a necessary step in quantum field theory of fundamental interactions. The fundamental structures, the gauge groups, which are the basis, for example, of the unified theory of electro-weak interactions are infinitely easier to deal with path integrals. And of course, this theory applies to an impressive number of other fields, in physics as well as in mathematics and engineering, as can be seen in Lawrence S. Schulman's book [27].

9.7.2 The Difficulty of Spin 1/2

In other words, Feynman achieved a sort of dream inscribed in Hamilton's reflections, that of having a unique theory to describe the physical world within well-defined and compatible limits. This dream had first been glimpsed by Dirac and many people took part in the adventure.

Is it really the case? Not quite. We stumble upon one of the most difficult subjects to unravel in the birth of quantum mechanics at the beginning of the 20th century: spin 1/2. This quantum physical quantity, which remained a mystery for more than twenty-five years between Zeeman's 1896 measurements of atomic level cleavages in even numbers and the final solution given by Uhlenbeck, Goudsmit and Pauli in 1925-26. The spin was an enigma because, precisely, it does not correspond to any intuitive notion of the world around us. And yet, it is a fundamental physical quantity, without which we would not understand the structure of matter. At the end of his book, Feynman wrote "The path integrals suffer from a terrible defect. They do not allow discussion of spin operators or other operators of this type in a simple and lucid way."

Indeed, spin is not an intuitive quantity! Of course, Feynman[16] has done this, attempting to harmonize integration formulas with quaternions, but the lack of commutativity of these numbers is a serious complication. The optimism of great minds is always fascinating.[17]

In addition, a current topic of great interest is *entangled states*, which are the basis of an impressive amount of research and technological discoveries in quantum information, including quantum cryptography and the Holy Grail of the quantum computer. But we do not see this topic appearing together with a label of path integral. In states called "GHZ",[18] there is not only a physical quantity that is out of our intuitive perception, but physical states whose measurement leads to paradoxical results in the same way as Schrödinger's cat.[19]

9.7.3 Optics and Analytical Mechanics

The previous considerations enlighten the relationship between analytical mechanics and optics. We have shown, in Sect. 5.1, Hamilton's remarkable statement in 1830 that the formalisms of optics and mechanics could be unified and that Newtonian

[16] R. P. Feynman, *An operator Calculus Having Applications in Quantum Electrodynamics*, Phys. Rev., vol 84, pp. 108–128, 1951.

[17] See also Lawrence Schulman, *A Path Integral for Spin*, Phys. Rev. 176, 1558, 1968; and the remarkable analysis of Alexander Altman, Ben D. Simons, *Condensed matter field Theory*, Cambridge University Press, (2010) Chap. 3, 134.

[18] D. M. Greenberger, M. Horne, and A. Zeilinger, in *Bell's Theorem, Quantum Theory, and Conceptions of the Universe*, edited by M. Kafatos, (Kluwer, Dordrecht, 1989); D. M. Greenberger, M. Horne, and A. Zeilinger *Going Beyond Bell's Theorem*, arXiv:0712.0921v1 [quant-ph] 2007; D. M. Greenberger, M. A. Horne, A. Shimony, and A. Zeilinger, *Bell's theorem without inequalities*, Am. J. Phys. **58**, 1131 (1990); N. D. Mermin, Phys. Today **43** (6).9 (1990).

[19] Jian-Wei Pan, D. Bouwmeester, M. Daniell, H. Weinfurter and A. Zeilinger, *Experimental test of quantum nonlocality in three photon GHZ entanglement*, Nature, **403** (6769), 515 (2000); Jian-Wei Pan and Anton Zeilinger, (2002) *Multi-Photo Entanglement and Quantum Non-Locality* https://vcq.quantum.at/fileadmin/Publications/2002-12.pdf.

mechanics corresponds to *the same limit* or approximation as geometrical optics as compared to wave optics.

In the same way as geometrical optics appears as the short wavelength limit of wave optics, classical mechanics appears as the limit of quantum mechanics for "small" values of $\hbar$, or rather in situations such that the value of the action is large compared with Planck's constant.

We have stressed in Sect. 9.5 the remarkable similarity between the Feynman principle in quantum mechanics and the Huygens–Fresnel principle that was the foundation of wave optics.

In Sect. 5.1, we studied the eikonal and how the transition between wave optics and geometrical optics is achieved, as well as the "semiclassical" approximation method of Wentzel, Kramers, and Brillouin. We shall not come back to this except by saying that several chapters of physics such as all optics and all mechanics therefore emerge as originating from the same common mold. (One thing remains to be done, and that is to extend this to field theory and its quantization, which is beyond the scope of this book.)

9.7.4 The Role of the Phase

In order to "see" completely how Feynman's principle emerges as a genuine variational principle, it would be necessary to treat the question of observables which can be found in the book by Feynman and Hibbs [25], Chap. 5, and more importantly, to explain the *variational principle of Schwinger*,[20] which can be found in [28].

Everything lies in the *phase*. One can make the naive and heuristic following remark. Consider a variation of the coordinates in (9.21) $x_i(t) \to x_i(t + \delta t)$ which leads to the variation δS of the action. Formally (i.e., without discussing precisely the mathematical nature of various expressions), we have a variation of the propagator

$$\delta K = \int_a^b \left(\delta \int_{t_a}^{t_b} \mathcal{L} dt \right) e^{\frac{i}{\hbar} S(b,a)} \mathcal{D} x(t),$$

(one can also vary the Lagrangian itself). The fact that the amplitude corresponding to the propagator is stationary to first order in the variation of the $x_i(t)$ implies the classical equation

$$\delta \int_{t_a}^{t_b} \mathcal{L} dt = 0;$$

i.e., the classical definition of the trajectory.

This can, of course, be made much more rigorous. The variational principle of Schwinger is perhaps less elegant than Feynman's, but it had undoubtedly better mathematical foundations when it was first formulated. One can refer to [25–28].

---

[20] Julian Schwinger, Phys. Rev. **82**, 914 (1951).

Feynman's principle can also be compared with what is called the *stationary phase* method in analysis. Consider the limit for $\mu \to \infty$ of the integral

$$G(\mu) = \int_{-\infty}^{\infty} e^{i\mu f(t)} \, dt. \tag{9.85}$$

For large values of μ, the phase of $e^{i\mu f(t)}$ varies very rapidly unless $f'(t) = 0$. Therefore, the dominant contributions to the integral will come from values of t for which $f'(t)$ vanishes. If $f'(t)$ vanishes at a single point t_0, we can expand f as a power series in the vicinity of t_0; i.e.,

$$G(\mu) = \int_{-\infty}^{\infty} e^{[i\mu(f(t_0)+\frac{1}{2}(t-t_0)^2 f''(t_0)+\cdots)]} \, dt. \tag{9.86}$$

If one neglects higher-order terms in the expansion, one obtains the result

$$G(\mu) = \sqrt{\frac{2\pi i}{\mu f''(t_0)}} e^{i\mu f(t_0)}, \tag{9.87}$$

to be compared with (9.31).

Feynman's principle is to take into account, in the calculation of an amplitude, the largest "number" of possible paths, with the *constraint* that paths too far from each other will lead to destructive interference. One can also visualize this as the fact that an amplitude increases when the "volume" of the space of the alternative paths that contribute in a coherent manner is larger. From that point of view, the phase of an amplitude acquires a physical role that perhaps is not fully appreciated. Both on the level of ideas and on the technical level, the method of Feynman's path integrals is of an undeniable depth and richness. It extends to many other physical problems: quantum field theory, but also Brownian motion, polarons, spin physics, statistical mechanics, critical phenomena etc. as we can see in Schulmann's book [27] where a pleasant discussion of quantum mechanics in curved spaces can be found.

9.8 Exercises

9.1. Propagator of a Harmonic Oscillator

The Lagrangian of a one-dimensional harmonic oscillator is

$$\mathcal{L} = \frac{1}{2}m\dot{x}^2 - \frac{1}{2}m\omega^2 x^2.$$

Show that the corresponding propagator is

$$K = F(T) \exp\left(\frac{im\omega}{2\hbar \sin \omega T}[(x_a^2 + x_b^2)\cos \omega T - 2x_a x_b]\right), \qquad (9.88)$$

where $T = t_b - t_a$ and where

$$F(T) = \left(\frac{m\omega}{2\pi i \hbar \sin \omega T}\right)^{1/2}.$$

Solutions of the Problems and Exercises

Exercises of Chapter 2

2.1. Kirschoff's Laws

The variational principle here consists in imposing that *the energy losses by Joule heating are as small as possible.* In other words, we want to find the minimum value of

$$W = R_1 I_1^2 + R_2 I_2^2 \quad \text{with the constraint} \quad I_1 + I_2 = I.$$

We find the zero of the derivative of $W = R_1 I_1^2 + R_2 (I - I_1)^2$ with respect to I_1, and this results in $R_1 I_1 = R_2 I_2$, which is of course the same as if we impose that the potential difference V between the two nodes is given. Notice that we do not need the notion of electric potential. We have replaced the *local* notion of potential difference by a *global energetic condition* and a very simple principle.

Considering an arbitrary circuit, the principle is that the global heating loss $\sum_k R_k I_k^2$ is minimal. Of course, one recovers the Kirchhoff laws. For a relatively simple network, the two approaches are equivalent. In practice, they may be very different if we consider a large electrical network of electricity, with, for instance, 10 million elements. Inverting a $10^7 \times 10^7$ matrix in real time is not realistic, whereas mathematical optimization procedures are extremely efficient and easy to handle.

2.2. Shape of a Massive String

Equilibrium corresponds to the configuration where the gravitational potential energy of the string is minimal. Consider an arbitrary shape of the string $z(x)$. An element of the string in the interval $[x, x + dx]$ has a length $dl^2 = dx^2 + dz^2 = (1 + \dot{z}(x)^2) dx^2$, and its potential energy is $dV = \mu g z \, dl$ (g is the acceleration of gravity). We must therefore minimize the integral

J.-L. Basdevant, *Variational Principles in Physics*,
https://doi.org/10.1007/978-3-031-21692-3

217

$$V = \int_0^a \mu g z \sqrt{1 + \dot{z}(x)^2} dx. \tag{A.1}$$

The Lagrange–Euler equation yields

$$1 + \dot{z}^2 = z\ddot{z}, \tag{A.2}$$

which we have already considered in (2.16). (This equation is frequently encountered in this kind of problem because it is one of the few cases where an analytic solution is available.) As we already know, the solution is

$$z = c \cosh((x - x_0)/c)$$

where the parameters c and x_0 are determined by the *constraints* $z(0) = z_0, z(a) = z_1$, and the length of the string $L = \int_0^a \sqrt{1 + \dot{z}(x)^2} dx$.

The minimum is located in the interval $x \in [0, a]$ according to the relative positions of the endpoints.

In this exercise one can see that by using the technique of Lagrange multipliers, which we defined in Sect. 2.4.3, the problem can be cast as a translation-invariant problem along the z axis since the length L is an intrinsic quantity of the string.

2.3. Soap Bubbles

Consider the interval $\{z, z + dz\}$ and let $r(z)$ be the radius of a transverse section of the surface. We want to minimize the energy

$$A = \int_{-h}^h 2\pi r \sqrt{1 + \dot{r}^2} dz$$

with the boundary conditions (constraints) $r(-h) = r(h) = R$. The problem is strictly the same as in the case of the string (A.1). The solution is

$$r = a \cosh(z/a), \quad \text{with} \quad R = a \cosh(h/a).$$

This surface, which is rotation invariant around the z axis, bears the sweet name of a catenoid (Fig. A.1).

One can attempt to determine shapes of bubbles attached to more complicated structures. (In general this must be done numerically.)

In exercise (2, 2) one can see that by using the technique of Lagrange multipliers, which we define in Sect. 2.4.3, the problem can be cast as a translation-invariant problem along the z axis since the length L is an intrinsic quantity of the string.

Fig. A.1 Soap bubble
between two symmetric
circles

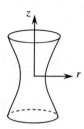

2.4. Conserved Quantities

We obtain directly $d\Gamma/dz = \dot{r}(1 + \dot{r}^2 - r\ddot{r})/(1 + \dot{r}^2)^{3/2}$. The equation of the curve
is $1 + \dot{r}^2 - r\ddot{r} = 0$, from which the result follows.

Therefore

$$r(z) = a\sqrt{1 + \dot{r}(z)^2}.$$

Setting $\dot{r}(z) = \sinh(\phi(z))$, we obtain

$$r(z) = a\cosh(\phi(z)); \quad \text{i.e.,} \quad \dot{r} = a\dot{\phi}(z)\sinh(\phi(z)),$$

and therefore $a\dot{\phi}(z) = 1$ and the solution $r(z) = a\cosh((z - z_0)/a)$. This is a particular case of the use of conserved quantities discussed in Chap. 3.

2.5. Lagrange Multipliers

We must minimize

$$V = \int_0^a \mu g z \sqrt{1 + \dot{z}(x)^2}\,dx, \tag{A.3}$$

with the constraints

$$z(0) = z_0, \quad z(a) = z_1, \quad \text{and} \int_A^B \sqrt{1 + \dot{z}(x)^2}\,dx = L.$$

One can transform the problem into

$$\min \bar{V} = \int_0^a (\mu g z + \lambda)\sqrt{1 + \dot{z}(x)^2}\,dx, \tag{A.4}$$

with $z(0) = z_0, \quad z(a) = z_1$.

The conserved quantity

$$\frac{(\mu g z + \lambda)}{\sqrt{1 + \dot{z}(x)^2}} = C \tag{A.5}$$

yields $\dot{z} = \sinh\phi(x)$, i.e., $\mu g z + \lambda = C\cosh\phi$ with $C\dot{\phi} = \mu g$. The solution is

$$z = -\frac{\lambda}{\mu g} + \frac{C}{\mu g} \cosh\left(\frac{\mu g}{C}(x - x_0)\right). \tag{A.6}$$

The constants x_0, C, and λ are fixed by the conditions $z(0) = z_0$, $z(a) = z_1$, and $\int_0^a \sqrt{1 + \dot{z}(x)^2}\,dx = L$.

2.6. Problem. Win a Downhill

1. With this definition of the variable x, we have $(z - z_0) = (x - x_0)\sin\alpha$ and the potential energy is $V = mg(z - z_0) = -mgx\sin\alpha$.
2. The total energy is $E = \frac{1}{2}m(\dot{x}^2 + \dot{y}^2) - mgx\sin\alpha$. Since energy is conserved, and since it is taken to be zero initially, we have $\dot{x}^2 + \dot{y}^2 = 2gx\sin\alpha$.
3. Therefore $dt^2 = (dx^2 + dy^2)/(2gx\sin\alpha)$.
4. The total time to get from O to A is therefore

$$T = \int_O^A dt = \frac{1}{\sqrt{2g\sin\alpha}} \int_0^A \sqrt{\frac{1 + (y')^2}{x}}\,dx$$

5. Using the Lagrange–Euler equation, we obtain

$$0 = \frac{d}{dx}\frac{y'}{\sqrt{x(1 + (y')^2)}}.$$

6. We deduce

$$C = \frac{y'}{\sqrt{x(1 + (y')^2)}},$$

where C is a constant. However,

$$\frac{y'}{\sqrt{x(1 + (y')^2)}} = \frac{dy}{\sqrt{x(dx^2 + dy^2)}} = \frac{\dot{y}}{\sqrt{x(\dot{x}^2 + \dot{y}^2)}} = \frac{\dot{y}}{x\sqrt{2g\sin\alpha}} = C, \tag{A.7}$$

and therefore $\dot{y} = Kx$ with $K = C\sqrt{2g\sin\alpha}$.
7. The parametric form $x(\theta) = (1 - \cos 2\theta)/2C^2 = \sin^2\theta/C^2$, $y(\theta) = (2\theta - \sin 2\theta)/2C^2$ satisfies the equation $(y')^2 = C^2x/(1 - C^2x)$; i.e., $(dy/d\theta)^2 = (dx/d\theta)^2 \tan^2\theta$. From $\dot{y}/x = K$, we obtain $(dy/d\theta)(d\theta/dt)/x = K$; i.e., $d\theta/dt = K/2$ and $\theta = Kt/2$ since, for $t = 0$, $\theta = 0$.
8. The curve is a portion of a cycloid. We have $dy/dx = \tan\theta$ and therefore $y' \gg 1$ for $\theta \sim \pi/2$. The trajectory starts vertically $(dy/dx = 0$ for $\theta = 0)$ and becomes horizontal if $y(A) \gg x(A)$, as shown in Fig. 7.1.
9. Since point A is fixed, the velocity v_A at A is fixed by energy conservation. It is the maximum velocity of the skier. Therefore, the time to get horizontally from $y(A)$ to $y(0)$ is larger than the time $(y(A) - y(0))/v_A$ it would take to cover this distance at the maximum velocity. On the other hand, one must start vertically in

Fig. A.2 Optimal trajectory
from O to A

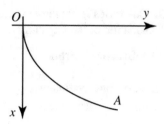

order to acquire the maximum velocity as quickly as possible. The ideal trajectory
comes from an optimization between these two effects (Fig. A.2).

Exercises of Chapter 3

3.1. Moving Pendulum

The Lagrangian is:

$$\mathcal{L} = \frac{m_1 + m_2}{2}\dot{x}^2 + \frac{m_2}{2}(l^2\dot{\phi}^2 + 2l\dot{x}\dot{\phi}\cos\phi) + m_2gl\cos\phi.$$

3.2. Properties of the Action

1. Free particle

$$S = \frac{m}{2}\frac{(x_2 - x_1)^2}{t_2 - t_1}.$$

2. Harmonic oscillator

$$S = \frac{m\omega}{2\sin\omega T}\left((x_2^2 + x_1^2)\cos\omega T - 2x_2x_1\right).$$

3. Constant force

$$S = \left(\frac{m}{2}v_0^2 - Fx_1\right)(t_2 - t_1) - \frac{1}{3}\left(\frac{F}{m}\right)^2(t_2 - t_1)^3$$

with $v_0 = (x_2 - x_1)/(t_2 - t_1) - (1/2)(F/m)(t_2 - t_1)$.
4. One varies the endpoint of arrival in the integration by parts of

$$\delta S = \int_{t_1}^{t_2}\left(\frac{\partial\mathcal{L}}{\partial x}\delta x(t) + \frac{\partial\mathcal{L}}{\partial\dot{x}}\delta\dot{x}(t)\right)dt.$$

5. One varies t_2, taking into account that the variation of the time of arrival yields a variation of the trajectory.

3.3. Brachistochrone

Energy conservation gives

$$\frac{1}{2}\left(\frac{ds}{dt}\right)^2 + g(z - \alpha) = 0. \tag{A.8}$$

We want to minimize

$$T = \int_a^b \left(\sqrt{\frac{1 + \dot{z}^2}{2g(\alpha - z)}}\right) dx \tag{A.9}$$

with the constraints $z(a) = \alpha$, $z(b) = \beta$.

The Lagrange function $\mathcal{L} = \sqrt{1 + \dot{z}^2/2g(\alpha - z)}$ does not depend on x, and therefore there is conservation of

$$\dot{z}\frac{\partial \mathcal{L}}{\partial \dot{z}} - \mathcal{L} = \frac{-1}{2g(1 + \dot{z}^2)(\alpha - z)} = -\frac{1}{\sqrt{R}}, \tag{A.10}$$

where we introduce a positive constant R. Setting $\dot{z} = \tan(\phi/2)$, we obtain the parametric form

$$z - z_0 = -\frac{R\cos\phi}{2}, \quad x - x_0 = \frac{R(\phi + \sin\phi)}{2}, \tag{A.11}$$

which is the equation of a cycloid.

3.4. Conjugate Momenta in Spherical Coordinates

1. The Lagrangian is $\mathcal{L} = \frac{1}{2}m(\dot{r}^2 + r^2\,\dot{\theta}^2 + r^2\sin^2\theta\,\dot{\phi}^2) - V(r)$.
2. The conjugate momenta are

$$p_r = \frac{\partial \mathcal{L}}{\partial \dot{r}} = m\dot{r}, \quad p_\theta = \frac{\partial \mathcal{L}}{\partial \dot{\theta}} = mr^2\dot{\theta}, \quad p_\phi = \frac{\partial \mathcal{L}}{\partial \dot{\phi}} = mr^2\sin^2\theta\dot{\phi}.$$

3. Taking the derivative of (3.84) with respect to time, and taking into account that in Cartesian coordinates $\mathbf{p} = m\mathbf{v}$, one obtains directly the result $L_z = mr^2\sin^2\theta\dot{\phi} = p_\phi$.
4. The conservation of p_ϕ, or L_z, corresponds to the invariance under translation in ϕ; i.e., rotation invariance around the z axis.
5. If a charged particle is in a magnetic field $\mathbf{B}$ parallel to Oz, there is rotational invariance around the z axis and the component L_z is conserved.

3.6. Problem. Strategy of a Regatta

1. We have by definition $\dot{x} = v_x = v\cos\theta$, $\dot{z} = v_z = v\sin\theta$, and therefore $z' = dz/dx = \tan\theta$.
2. We have $v_x = v\cos\theta = w/h$. This velocity is maximum when $h(z')$ is minimum; i.e., for $z' = 1$, namely $\theta = \pi/4$. We then have $v_x = w/2$. In fact, it is sufficient to multiply h by a constant to be in the appropriate situation for a given sailboat for which $v_{x,\max} = \lambda w$.
3. We have $dt = dx/v_x = h'(z')\,dx/w(z)$, and therefore

$$T = \int_0^L dx\,\frac{h'(z')}{w(z)}. \tag{A.12}$$

4. Setting $\Phi = h'(z')/w(z)$, the Lagrange–Euler equation that optimizes the total time T is

$$\frac{\partial\Phi}{\partial z} = \frac{d}{dx}\left(\frac{\partial\Phi}{\partial z'}\right).$$

5. The function Φ does not depend explicitly on x. Therefore, we have

$$\frac{d}{dx}\Phi = z'\frac{\partial\Phi}{\partial z} + z''\frac{\partial\Phi}{\partial z'}.$$

Consequently,

$$\frac{d}{dx}\left(\Phi - z'\frac{\partial\Phi}{\partial z'}\right) = 0,$$

which gives $(h'(z')z' - h(z'))/w(z) = \text{constant}$.

6. We have $z'h' - h = -2/z'$. We therefore obtain the first-order differential equation for the function $x(z)$, $(-2/A)dx/dz = w(z)$, and hence the result

$$x = L\frac{w_0 z - w_1 z_0 \ln(1 + (z/z_0))}{w_0 z_1 - w_1 z_0 \ln(1 + (z_1/z_0))}, \tag{A.13}$$

where we have incorporated the conditions $(x = 0, z = 0)$ and $(x = L, z = z_1)$.

7. We obtain

$$z' = \frac{dz}{dx} = \frac{w_0 z_1 - w_1 z_0 \ln(1 + (z_1/z_0))}{w_0 L - w_1 L z_0/(z + z_0)}.$$

If $z_1 \ll L$ and $z_1 \ll z_0$, the velocity of the wind does not vary appreciably over the whole path, and one has $z' \sim z_1/L \ll 1$.

In the second question, we have seen that the optimal velocity for a constant wind velocity is attained for $z' = 1$. The present configuration certainly does not correspond to the best strategy. One must tack at some point (x_1, Z) with $0 < x_1 < L$

Fig. A.3 Path of the boat
with a tacking at $x = L/2$

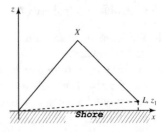

and $Z \gg z_1$, as represented in Fig. A.3 in order to benefit fully from the power of
the wind (this possibility was excluded in the text).

The trajectory drawn with an angle of $\theta = 45$ degrees ($|z'| = 1$) and a tacking $\theta \rightarrow$
$-\theta$ at $x = L/2$ has a total length $L\sqrt{2}$ and a velocity greater than $(w0 - w1)/2$. The
time along this path, $T_v = 2L\sqrt{2}/(w0 - w1)$, is obviously shorter than the time
along the path with no tacking, $T \sim 2L(z_1/L)/(w0 - w1) = 2z_1/(w0 - w1)$.

In realistic cases, for instance the America's Cup, one can see how subtle the
strategy of a regatta problem is. Skippers must make quick decisive choices between
very different options.

Exercises of Chapter 4

4.1. Poisson Brackets of Angular Momentum

The Poisson brackets are:

$$\{L_x, L_y\} = L_z \qquad \{L_y, L_z\} = L_x \qquad \{L_z, L_x\} = L_y$$

4.2. The Lenz Vector

We obtain

$$\{A_x, A_y\} = -\frac{2}{m}HL_z \, , \{A_x, L_x\} = 0 \, , \{A_x, L_y\} = A_z \, ,$$

and cyclic permutations. Besides that its Poisson bracket with the hamiltonian vanishes. This leads to the impression of 7 conserved quantities in the Kepler or Hydrogen atom problems (two vectors and the energy). Actually these reduce to 5, given
that $\mathbf{L} \cdot \mathbf{A} = 0$ (that vector lies in the plane of the orbit, and furthermore there is
a scalar relation between it, the coupling constant and the energy). This is called
"superintegrability", since there are only 6 physical quantities in the problem: $(\mathbf{r}, \mathbf{p})$.

The Runge-Lenz vector has played an incredible role in the history of mechanics.[1] Actually, in Newtonian dynamics, (Potential k/r), the Lenz vector points to
the perihelion of planets. Einstein remarked that about the planet Mercury, since in

---

[1] H. Goldstein, *Prehistory of the Runge-Lenz Vector*, American Journal of Physics, **43**, 735 (1975).

general relativity there is a slight (an correct) deviation from that.

4.3. Three Coupled Oscillators

Let us recall the result for two oscillators.

1. One obtains directly

$$\{X, P\} = 1 \quad \{X, Q\} = 0 \quad \{Y, P\} = 0 \quad \{Y, Q\} = 1$$

$$H = \frac{P^2}{2m} + \frac{m\omega^2 X^2}{2} + \frac{Q^2}{2m} + \frac{m(\omega^2 + \Omega^2)Y^2}{2}.$$

2. The eigenfrequencies of the system are therefore $\omega_1 = \omega$ and $\omega_2 = \sqrt{\omega^2 + \Omega^2}$.
3. The general form of the motion follows from

$$X = A\cos(\omega_1 t + \phi), \quad Y = B\cos(\omega_2 t + \psi).$$

In the case of three coupled oscillators, we start from

$$H = \frac{m}{2}(p_1^2 + p_2^2 + p_3^2) + \frac{m\omega^2}{2}(x_1^2 + x_2^2 + x_3^2) + \frac{3m\Omega^2}{2}(x_1^2 + x_2^2).$$

The canonical transformation (Jacobi variables) gives:
$$X_1 = (x_1 - x_2)/\sqrt{2}, \quad X_2 = (2x_3 - x_1 - x_2)/\sqrt{6}, \quad X_3 = (x_1 + x_2 + x_3)/\sqrt{3}$$
$$P_1 = (p_1 - p_2)/\sqrt{2}, \quad P_2 = (2p_3 - p_1 - p_2)/\sqrt{6}, \quad P_3 = (p_1 + p_2 + p_3)/\sqrt{3}$$

From which, we obtain:
$$\{X_i, P_j\} = \delta_{ij} \quad .$$

Hence the Hamiltonian, sum of three independent Hamiltonians,

$$H = \sum_i ((P_i^2 + \omega_i^2 X_i^2)/2), \quad \text{with} \quad \omega_1 = \omega_2 = \sqrt{\omega^2 + 3\Omega^2} \quad \omega_3 = \omega \quad .$$

The "path" in phase space of these three circular motions is not easy to mimic on a two-dimensional graph. Notice that the mode for which the initial three oscillators are in phase is done at the initial pulsation ω, which is not guessed easily on the formulation $(x, \dot{x})$.

4.4. Forced Oscillations

1. We obtain with no difficulty

$$H = \frac{\omega}{2}(X^2 + P^2), \quad \{X, P\} = 1.$$

2. In these variables, which are the same as those used by Dirac in the quantum harmonic oscillator,

$$H = \omega(a^* a).$$

3. We obtain $\{a, a^*\} = -i$.
4. The evolution equation in time of a is

$$\dot{a} = \{a, H\} = -i\omega a,$$

which is a first-order differential equation. The general solution is

$$a(t) = a_0 \exp(-i\omega t),$$

where a_0 is a complex constant. The energy of the oscillator is $E = \omega|a_0|^2$.
5. For $t \leq 0$, we have $a_0 = 0$. In the presence of H_{pot}, the Hamiltonian becomes

$$H = \omega(a^* a) + b(a + a^*) \sin \Omega t.$$

Therefore, we have

$$\dot{a} = \{a, H\} = -i\omega a - ib \sin \Omega t.$$

This is solved by standard techniques. With the condition $E(t < 0) = 0$, one obtains

$$E(t > T) = \omega b^2 \left| \frac{e^{-i(\Omega-\omega)T} - 1}{2i(\Omega - \omega)} + \frac{e^{-i(\Omega+\omega)T} - 1}{2i(\Omega + \omega)} \right|^2.$$

6. This is a *resonance* phenomenon at $\Omega = \omega$ (or at $\Omega = -\omega$, which is equivalent). In the vicinity of $\Omega = \omega$, the energy acquired by the oscillator is of the form

$$E(t > T) = \omega b^2 \frac{\sin^2(\Omega - \omega)T/2}{(\Omega - \omega)^2},$$

which has a peak of height $\omega b^2 T^2/4$ at $\Omega = \omega$.

4.8. Problem

Closed Chain of Coupled Oscillators

1. (a) In the definition, we see that

$$y_k = y_{N-k}^*, \quad q_k = q_{N-k}^*.$$

(b) We have

$$\sum_{k=1}^{N} y_k y_k^* = \sum_{k=1}^{N} \left(\frac{1}{\sqrt{N}} \sum_{n=1}^{N} e^{2ikn\pi/N} x_n \right) \left(\frac{1}{\sqrt{N}} \sum_{n'=1}^{N} e^{-2ikn'\pi/N} x_n' \right). \quad \text{(A.14)}$$

The summation over k gives $\delta_{nn'}$ and the result

$$\sum_{k=1}^{N} q_k q_k^* = \sum_{n=1}^{N} p_n^2. \quad \text{(A.15)}$$

Similarly

$$\sum_{k=1}^{N} q_k q_k^* = \sum_{k=1}^{N} \left(\frac{1}{\sqrt{N}} \sum_{n=1}^{N} e^{-2ikn\pi/N} p_n \right) \left(\frac{1}{\sqrt{N}} \sum_{n'=1}^{N} e^{2ikn'\pi/N} p_n' \right). \quad \text{(A.16)}$$

The summation over k gives $\delta_{nn'}$, and hence the result.

(c) On the other hand, we have

$$\sum_{n=1}^{N} (x_n - x_{n+1})^2 = \frac{1}{N} \sum_{n=1}^{N} \left(\sum_{k=1}^{N} e^{-2ikn\pi/N} (1 - e^{-2ik\pi/N}) y_k \right)$$
$$\times \left(\sum_{k'=1}^{N} e^{2ik'n\pi/N} \left(1 - e^{2ik'\pi/N} \right) y_{k'}^* \right). \quad \text{(A.17)}$$

The summation over n gives $\delta_{kk'}$ and the result.

2. Equations of motion and their solution.

(a) We have

$$H = \sum_{k=1}^{N} \left[\frac{q_k q_k^*}{2m} + \frac{1}{2} m \Omega'^2_k y_k y_k^* \right]$$

with

$$\Omega'^2_k = \omega^2 + 4\Omega^2 \sin^2 \left(\frac{k\pi}{N} \right).$$

(b) We have

$$\{y_j, q_k\} = \delta_{jk}, \{y_j^*, q_k^*\} = \delta_{jk}, \{y_j, q_{N-k}^*\} = \delta_{jk}, \{y_j^*, q_{N-k}\} = \delta_{jk}. \quad \text{(A.18)}$$

(c) We obtain

$$\dot{y}_k = \{y_k, H\} = \frac{m}{2}(q_k^* + q_{N-k}) = mq_k^*,$$

$$\dot{y}_k^* = \{y_k^*, H\} = \frac{m}{2}(q_k + q_{N-k}^*) = mq_k,$$

$$\dot{q}_k = \{q_k, H\} = -\frac{m\Omega_k'^2(y_k^* + y_{N-k})}{2} = m\Omega_k'^2 y_k^*,$$

$$\dot{q}_k^* = \{q_k^*, H\} = -\frac{m\Omega_k'^2(y_k + y_{N-k}^*)}{2} = m\Omega_k'^2 y_k.$$

(d) We therefore have $\{y_k(t)\} = a_k \cos(\Omega'_k t + \phi_k)$, and hence $\{x_n(t)\}$.

3. If, at time $t = 0$, we have $y_N(0) = 1$, $\dot{y}_N(0) = 0$ and $\{y_n(0) = 0, \dot{y}_n(0) = 0\}$, $\forall n \neq N$, then $y_N(t) = \cos(\omega t)$ and $y_n(t) = 0, \forall n \neq N$. Therefore $x_n(t) = (1/\sqrt{N})$ $\cos(\omega t)$. Oscillators of the same amplitude at a given time are always in phase, and only the global motion with respect to the plane $x = 0$ with frequency ω appears.

4. Wave propagation.
   If $\omega = 0$, the eigenfrequencies are $\Omega'_k = 2\Omega \sin(k\pi/N) \sim 2\Omega(k\pi/N)$ for $k \ll N$. The boundary conditions give $y_1 = \cos 2\Omega\pi t/N$, $y_{N-1} = \cos 2\Omega\pi t/N$, and $y_n = 0$ otherwise.

   (a) Therefore, we obtain

   $$lx_n = x_{N-n} = \frac{2}{\sqrt{N}} \cos\left(\frac{2\Omega\pi t}{N}\right) \cos 2n\pi N \tag{A.19}$$

   $$= \frac{1}{\sqrt{N}}\left[\cos\left(\frac{2\Omega\pi t + 2n\pi}{N}\right) + \cos\left(\frac{2\Omega\pi t - 2n\pi}{N}\right)\right]. \tag{A.20}$$

   (b) We observe a propagation phenomenon in both directions since

   $$x_{n+m}(t) = x_n(t \pm m/\Omega)$$

   in the notation above. The point x_{n+m} has the same amplitude at time $t + m/\Omega$ as the point x_n at time t.

   (c) If we write $x_n(t) = f(t, y = na)$, the function f is

   $$f(t, y) = \frac{1}{\sqrt{N}}\left[\cos\left(\frac{2\Omega\pi t + 2y\pi/a}{N}\right) + \cos\left(\frac{2\Omega\pi t - 2y\pi/a}{N}\right)\right]$$

   and satisfies the wave equation

   $$\frac{1}{\Omega^2 a^2}\frac{\partial^2 f}{\partial t^2} - \frac{\partial^2 f}{\partial x^2} = 0.$$

In this chain of coupled oscillators, a progressive wave of velocity Ωa propagates.

Exercises of Chapter 5

5.1. The Lorentz Hamiltonian

The proof takes some time, but it is straightforward (follow Sect. (3.3.2)).

5.2. Virial Theorem

1. One obtains

$$\{A, H\} = \frac{p^2}{m} - \mathbf{r} \cdot \nabla V.$$

The time evolution of A is simply

$$\frac{dA}{dt} = \{A, H\} = \frac{p^2}{m} - \mathbf{r} \cdot \nabla V.$$

2. We have $\langle \dot{A} \rangle = (A(T) - A(0))/T = 0$. Therefore, inserting this in the result above, we obtain

$$2\left\langle \frac{p^2}{2m} \right\rangle = \langle \mathbf{r} \cdot \nabla V \rangle.$$

3. If $V = g\, r^n$, we have

$$\mathbf{r} \cdot \nabla V = r \frac{\partial V}{\partial r} = n V.$$

We therefore obtain $2\langle E_c \rangle = n\langle V \rangle$.
4. The total energy is $E = E_c + V$. We therefore obtain
   (a) For a harmonic oscillator, $E = 2\langle E_c \rangle = 2\langle V \rangle$.
   (b) For a Newtonian potential, $E = -\langle E_c \rangle = (1/2)\langle V \rangle$, which is obvious on a circular trajectory, but holds for any elliptic trajectory.
5. In general, for an arbitrary potential, the orbits of bound states are not closed. However, they remain confined in a given region of space at any time. The generalization of the averaging (5.62) is

$$\langle f \rangle = \lim_{(T \to \infty)} \frac{1}{T} \int_0^T f(t)\, dt.$$

With this definition, we have

$$\langle \dot{A} \rangle = \lim_{(T \to \infty)} (A(T) - A(0))/T = 0$$

since $A(t)$ is bounded for any t. With this definition, the result remains true.

Problem of Chapter 6. Neutron Transport in Matter

The Lagrangian density is

$$L = \frac{3}{v^2} \frac{\partial \psi}{\partial t} \frac{\partial \psi^*}{\partial t} - \nabla \psi \cdot \nabla \psi^* - \frac{a^2}{2} \left(\psi^* \frac{\partial \psi}{\partial t} - \psi \frac{\partial \psi^*}{\partial t} \right), \tag{A.21}$$

where ψ^* is the "mirror" density which concentrates instead of diffusing. This leads to the propagation equation

$$\frac{3}{v^2} \frac{\partial^2 \psi}{\partial t^2} - \Delta \psi + a^2 \frac{\partial \psi}{\partial t} = 0. \tag{A.22}$$

This equation can be solved by Fourier transformation if the coefficients v and a are constants. (This is not the case if the medium is inhomogeneous or discontinuous.)

Exercises of Chapter 7

7.1. Geodesics

Solutions exist only for $\rho \geq R$ (which is explained by Eq. (7.125)).
   The energy is

$$E = \frac{m}{2} \left(\dot{\rho}^2 \frac{R^2}{\rho^2 - R^2} + \frac{A^2}{\rho^2} \right). \tag{A.23}$$

The calculation is similar to previous cases such as (2). We define the parameters ω and γ as before:

$$\omega^2 = \frac{2E}{mR^2}, \quad \gamma^2 = \frac{mA^2}{2ER^2}. \tag{A.24}$$

We obtain

$$\rho(t) = R\sqrt{1 + (1 - \gamma^2) \sinh^2 \omega(t - t_0)}, \tag{A.25}$$

and

$$\tanh(\phi(t) - \phi_0) = \gamma \tanh \omega(t - t_0). \tag{A.26}$$

7.2. Hyperbolic Geodesics

Consider the metric

$$ds^2 = \frac{R^2}{\rho^2 + R^2} \, d\rho^2 + \rho^2 \, d\theta^2 + \rho^2 \sin^2 \theta \, d\phi^2, \tag{A.27}$$

where R is a positive characteristic length.

Notice that one considers this metric as deriving from a "Lorentzian" metric, if one changes the sign of R

$$ds^2 = dx^2 + dy^2 + dz^2 - dw^2, \tag{A.28}$$

by the three-dimensional reduction

$$w^2 = x^2 + y^2 + z^2 + R^2 = \rho^2 + R^2. \tag{A.29}$$

The calculation is very similar to that of Example 2.

In spherical coordinates, the Lagrangian of the problem is

$$\mathcal{L} = \frac{m}{2} \left(\dot{\rho}^2 \left(\frac{R^2}{\rho^2 + R^2} \right) + \rho^2 \dot{\theta}^2 + \rho^2 \sin^2 \theta \dot{\phi}^2 \right). \tag{A.30}$$

The conservation laws are the same as before.

1. Rotation invariance yields the conservation of angular momentum. The motion is planar.
2. Choosing the direction of the angular momentum as the polar axis, we have $\theta = \pi/2$ and $\dot{\theta} = 0$.
3. The Lagrangian of the planar motion reduces to

$$\mathcal{L} = \frac{m}{2} \left(\dot{\rho}^2 \left(\frac{R^2}{\rho^2 + R^2} \right) + \rho^2 \dot{\phi}^2 \right). \tag{A.31}$$

4. The conservation of angular momentum leads to

$$\frac{d}{dt}(\rho^2 \dot{\phi}) = 0 \implies \dot{\phi} = \frac{A}{\rho^2} \tag{A.32}$$

where A is a constant of the motion.

1. The energy, which is a constant of the motion, is

$$E = \frac{m}{2} \left(\dot{\rho}^2 \frac{R^2}{\rho^2 + R^2} + \frac{A^2}{\rho^2} \right). \tag{A.33}$$

2. The solution of the problem is obtained rather easily. One defines the parameters ω and γ as:

$$\omega^2 = \frac{2E}{mR^2} \quad \text{and} \quad \gamma^2 = \frac{mA^2}{2ER^2}. \tag{A.34}$$

The result is in a way similar to Example 2, except that hyperbolic functions replace some trigonometric functions,

$$\rho(t) = R\sqrt{\gamma^2 \cosh^2 w(t - t_0) + \sinh^2 w(t - t_0)} \qquad (A.35)$$

and

$$\tan(\phi(t) - \phi_0) = \gamma \coth w(t - t_0). \qquad (A.36)$$

We notice that the distance to the origin increases *exponentially* when $|t| \to \infty$. The geodesics of the metric (A.27) are hyperbolas

$$x^2 - y^2/\gamma^2 = R^2. \qquad (A.37)$$

Problem of Chapter 7. Free Motion on the Sphere S^3

We consider the case of free motion on the three-dimensional "spherical" space of a sphere S^3 imbedded in $\mathcal{R}^4$. Obviously, the volume of this space is finite since $\rho^2 = x^2 + y^2 + z^2 \leq R^2$.

In spherical coordinates, the Lagrangian of the problem is

$$\mathcal{L} = \frac{m}{2} \left(\dot{\rho}^2 \left(\frac{R^2}{R^2 - \rho^2} \right) + \rho^2 \dot{\theta}^2 + \rho^2 \sin^2 \theta \dot{\phi}^2 \right). \qquad (A.38)$$

The conservation laws of the problem which bring simplifications to the motion are

1. There is rotational invariance. The angular momentum is conserved, and the motion occurs on a plane.
2. We can choose the direction of the angular momentum as polar axis; i.e., $\theta = \pi/2$ and $\dot{\theta} = 0$.
3. The Lagrangian of the planar motion therefore reduces to

$$\mathcal{L} = \frac{m}{2} \left(\dot{\rho}^2 \left(\frac{R^2}{R^2 - \rho^2} \right) + \rho^2 \dot{\phi}^2 \right). \qquad (A.39)$$

4. The conservation of angular momentum, second Kepler's law, results in

$$\frac{d}{dt}(\rho^2 \dot{\phi}) = 0 \implies \dot{\phi} = \frac{A}{\rho^2}, \qquad (A.40)$$

where A is a constant, fixed by the initial conditions.
5. The energy, which is a constant of the motion, is therefore

$$E = \frac{m}{2} \left(\dot{\rho}^2 \frac{R^2}{R^2 - \rho^2} + \frac{A^2}{\rho^2} \right). \qquad (A.41)$$

The two constants of the motion E and A satisfy the inequality

$$A^2 \leq \frac{2R^2 E}{m}, \tag{A.42}$$

which is a direct consequence of the fact that the energy is greater than the rotational energy $mA^2/2\rho^2$. This is a consequence of (A.41); i.e., $E \geq mA^2/2\rho^2 \geq mA^2/2R^2$.

The Eqs. (A.40) and (A.41) are first-order differential equations that determine the motion in terms of the constants of the motion E and A.

The solution is simple. We define parameters ω and γ by

$$\omega^2 = \frac{2E}{mR^2} \quad \text{and} \quad \gamma^2 = \frac{mA^2}{2ER^2}. \tag{A.43}$$

From (A.42), we have the inequality

$$\gamma^2 \leq 1. \tag{A.44}$$

Setting

$$\rho = R\cos(\omega\psi); \quad \text{i.e.,} \quad \dot{\rho} = -\omega\dot{\psi}R\sin(\omega\psi). \tag{A.45}$$

If we insert this in Eq. (A.41), we obtain

$$\omega^2 = \omega^2\dot{\psi}^2 + \frac{\omega^2\gamma^2}{\cos^2(\omega\psi)}; \tag{A.46}$$

i.e.,

$$\omega^2\dot{\psi}^2 \cos^2(\omega\psi) = \omega^2 (\cos^2(\omega\psi) - \gamma^2). \tag{A.47}$$

We now make the change of functions

$$\sin(\omega\psi(t)) = \sqrt{1 - \gamma^2}\, u(t); \quad \text{therefore } \cos^2(\omega\psi) = 1 - (1 - \gamma^2)u^2. \tag{A.48}$$

Show that the choice

$$u(t) = \sin(\omega\zeta(t)) \tag{A.49}$$

leads with no difficulty to:

$$\dot{\zeta}^2 = 1, \quad \text{namely} \quad u = \sin(\omega(t - t_0)), \tag{A.50}$$

and to the result, i.e. The expressions of $\rho(t)$, $\tan(\phi(t) - \phi_0)$ as well as the frequency ω.

$$\rho(t) = R\sqrt{\cos^2 \omega(t - t_0) + \gamma^2 \sin^2 \omega(t - t_0)}, \qquad (A.51)$$

which is periodic and of frequency ω. The calculation of the time evolution of the azimuthal angle $\phi(t)$ is obtained by this expression and Eq. (A.40),

$$\dot{\phi} = \frac{A}{R^2(\cos^2 \omega(t - t_0) + \gamma^2 \sin^2 \omega(t - t_0))}; \qquad (A.52)$$

i.e.,

$$\tan(\phi(t) - \phi_0) = \gamma \tan \omega(t - t_0), \qquad (A.53)$$

which is also periodic and of frequency ω.

We conclude the following.

1. Consider the *Euclidean* plane of the motion (i.e., $x = \rho \cos \phi$, $y = \rho \sin \phi$). For simplicity, we choose the initial parameters as $t_0 = 0$, $\phi_0 = 0$, and we have

$$x = R \cos \omega t \qquad y = \gamma R \sin \omega t.$$

The trajectory is an ellipse of equation $x^2 + y^2/\gamma^2 = R^2$.

2. The point $\rho = R$ (i.e., the boundary of the space) is *always* reached, whatever the initial conditions on the energy and the angular momentum. If $A = 0$, the angular momentum vanishes and the motion is linear and sinusoidal. If $\gamma = 1$, the motion is uniform on a circle of radius R.

3. In this Euclidean plane, the energy of the particle is

$$E = \frac{1}{2}m(\dot{x}^2 + \dot{y}^2) + V,$$

where the "effective potential" V is energy dependent:

$$V = \frac{1}{2}m \frac{\rho^2 \dot{\rho}^2}{R^2 - \rho^2} = \frac{E\rho^2}{R^2}.$$

Therefore, the motion *also* appears as a two-dimensional harmonic motion whose frequency Ω depends on the *total energy* E,

$$\Omega^2 = \frac{2E}{mR^2}.$$

4. Of course, if the square of the velocity is a constant in the *curved* four-dimensional space, this is not the case if one visualizes the phenomenon in a Euclidean plane, as above.

5. The simplicity of the result is intuitive. Quite obviously, as one can see in the definition, the symmetry of the problem is much larger than the sole rotation in $\mathcal{R}^3$. There is a rotation invariance in $\mathcal{R}^4$. The solutions of maximal symmetry correspond to a

uniform motion on a circle of radius R in a plane whose orientation is arbitrary in $\mathcal{R}^4$. The whole set of solutions is obtained by projecting these particular solutions on planes of $\mathcal{R}^3$, which leads to the elliptic trajectories that we have found.

Exercises on Chapter 9

9.1. Propagator of a Harmonic Oscillator

Propagator of a Harmonic Oscillator

The classical action for a harmonic one-dimensional oscillator is

$$S_{cl} = \frac{m\omega}{2 \sin \omega T} (x_a^2 + x_b^2) \cos \omega T - 2x_a x_b.$$

The calculation of the propagator involves only Gaussian integrals, and the result follows directly.

References

1. L. Landau, E. Lifshitz, *Mechanics* (Pergamon Press, Oxford, 1965)
2. L. Landau, E. Lifshitz, *The Classical Theory of Fields* (Pergamon Press, Oxford, 1965)
3. H. Goldstein, C. Poole, J. Safko, *Classical Mechanics* (Addison Wesley, Boston, 2002)
4. R.P. Feynman, R.B. Leighton, M. Sands, *The Feynman Lectures on Physics* (Addison-Wesley, Reading MA, 1964)
5. P. Samuelson, *Foundations of Economic Analysis* (Harvard University Press, 1947), (Enlarged ed., 1983)
6. W. Yourgrau, S. Mandelstam, *Variational Principles in Dynamics and Quantum Theory* (Dover Publications, New York, 1979)
7. A. Koestler, *The Act of Creation* (Hutchinson & Co., London, 1964)
8. I.M. Gelfand, S. Vasilevich Fomin, *Calculus of Variations*, (Rev. English ed. Prentice-Hall, Englewood Cliffs, NJ, 1963). A.R. Forsyth, *Calculus of Variations* (Dover, New York 1960). J-P. Bourguignon, *Calcul Variationnel* (Ecole Polytechnique, Palaiseau 1990)
9. E. Schrödinger, *Statistical Thermodynamics* (Dover Publications, New York, 1989)
10. Y. Kosmann-Schwartzbach, *Les Théorèmes de Noether, Invariance et lois de conservation au XXe siècle*, Éditions de l'École Polytechnique, (2006). *The Noether Theorems: Invariance and Conservation Laws in the Twentieth Century*. Translated by Bertram Schwarzbach. Springer 2011, ISBN 978-0387878676
11. J.-L. Basdevant, J. Dalibard, *Quantum Mechanics* (Springer Verlag, Heidelberg, 2005)
12. M. Philip, *Morse and Herman Feshbach, Methods of Theoretical Physics* (Mc Graw-Hill, New York, 1953)
13. I. Percival, D. Richards, *Introduction to Dynamics* (Cambridge University Press, Cambridge, 1982)
14. M. Born, E. Wolf, *Principles of Optics* (Pergamon Press, Oxford, 1964)
15. A. Messiah, *Quantum Mechanics* (North-Holland, Amsterdam, 1962)
16. J.L. Basdevant, J. Rich, M. Spiro, *Fundamentals in Nuclear Physics* (Springer, New York, 2005)
17. D. Langlois, *General Relativity* (Vuibert, 2013)
18. N. Deruelle, J.-P. Uzan, *Theories of Relativity* (Belin, 2019)
19. E. Gourgoulhon, *General Relativity, Observatoire de Paris-Meudon*, http://luth.obspm.fr/~luthier/gourgoulhon/fr/master/relat.html (2013–2014)
20. H. Stefani, *General Relativity* (Cambridge University Press, Cambridge, 1982)

© The Editor(s) (if applicable) and The Author(s), under exclusive license to
Springer Nature Switzerland AG 2023
J.-L. Basdevant, *Variational Principles in Physics*,
https://doi.org/10.1007/978-3-031-21692-3

21. S. Weinberg, *Gravitation and Cosmology* (John Wiley & Sons, New York, 1972)

22. P.A.M. Dirac, *General Theory of Relativity* (John Wiley & Sons, New York, 1975)

23. C.W. Misner, K.S. Thorne, J.A. Wheeler, *Gravitation* (W.H. Freemann and Company, New York, 1973)

24. J. Rich, *Fundamentals of Cosmology* (Springer-Verlag, Heidelberg, 2001)

25. R.P. Feynman, A.R. Hibbs, *Quantum Mechanics and Path Integrals* (McGraw-Hill, New York, 1965)

26. L. Brown, *Feynman's Thesis, A new approach to Quantum Theory* (World Scientific, Singapore, 2005); Contains Feynman's Thesis at Princeton *The Principle of Least Action in Quantum Mechanics* (1942); R.P. Feynman *Space-Time Approach to Non-relativistic Quantum Mechanics*, Rev. Mod. Phys. 20,367 (1948); and P.A.M. Dirac *The Lagrangian in Quantum Mechanics* Physikalische Zeitschrift der Sowjetunion (1933)

27. S. Lawrence, *Schulman, Techniques and Applications of Path Integration* (John Wiley & Sons, New York, 1981)

28. J. Schwinger, *Selected Papers on Quantum Electrodynamics* (Dover, New York, 1958)

29. S. Albeverio, R. Hoegh-Krohn, S. Mazzuchi, *Mathematical Theory of Feynman Path Integrals*, Lecture Notes in Mathematics, vol. 523 (Springer-Verlag, 1976)

Index

© The Editor(s) (if applicable) and The Author(s), under exclusive license to
Springer Nature Switzerland AG 2023
J.-L. Basdevant, *Variational Principles in Physics*,
https://doi.org/10.1007/978-3-031-21692-3

Printed in the United States
by Baker & Taylor Publisher Services